*Springer Monographs in Mathematics*

T0202843

J. Coates · R. Sujatha

# Cyclotomic Fields and Zeta Values

J. Coates
Centre for Mathematical Sciences
DPMMS
Wilberforce Road
Cambridge, CB3 0WB, England
e-mail: J.H.Coates@dpmms.cam.ac.uk

R. Sujatha
School of Mathematics
Tata Institute of Fundamental Research
Homi Bhabha Road, Colaba
Mumbai 400 005, India
e-mail: sujatha@math.tifr.res.in

Mathematics Subject Classification (2000): 11R18, 11R23

ISSN 1439-7382

ISBN 978-3-642-06959-8        e-ISBN 978-3-540-33069-1

Springer is a part of Springer Science+Business Media
springer.com
© Springer-Verlag Berlin Heidelberg 2006
Softcover reprint of the hardcover 1st edition 2006

Cover design: Erich Kirchner, Heidelberg, Germany

# Preface

| Chihayaburu | *Mighty they are* |
|---|---|
| Kami no igaki ni | *The gods within this sacred shrine-* |
| Hau kuzu mo | *Yet even the vines* |
| Aki ni wa aezu | *Creeping in the precincts could not hold* |
| Utsuroinikeri | *Against the autumn's tingeing of* |
| | *their leaves.* |

– Ki no Tsurayaki (Kokinshu, V : 262).

This little book is intended for graduate students and the non-expert in Iwasawa theory. Its aim is to present in full detail the simplest proof of the important theorem on cyclotomic fields, which is often called "the main conjecture". We have thought it worthwhile to write such a book, not only because this theorem is arguably the deepest and most beautiful known result about the arithmetic of cyclotomic fields, but also because it is the simplest example of a vast array of subsequent, unproven "main conjectures" in modern arithmetic geometry involving the arithmetic behaviour of motives over $p$-adic Lie extensions of number fields (see [CFKSV]). These main conjectures are concerned with what one might loosely call the exact formulae of number theory which conjecturally link the special values of zeta and $L$-functions to purely arithmetic expressions (the most celebrated example being the conjecture of Birch and Swinnerton-Dyer for elliptic curves).

The first complete proof of the cyclotomic main conjecture was given by Mazur-Wiles, but it should not be forgotten that Iwasawa himself not only discovered the main conjecture but proved an important theorem which implies it in all known numerical cases. In this book, we follow this approach to the main conjecture via Iwasawa's theorem, and complete its proof by the ingenious arguments using Euler systems, due to Kolyvagin, Rubin and Thaine. Not only does this treatment have the advantage of using less machinery, but it also gives for example, a very simple proof of the existence of the $p$-adic analogue of the Riemann zeta function.

If one looks at the past evolution of algebraic number theory, there has been a tendency to discover that the ideas which initially seem special to cyclotomic fields do, in the end, turn out to have very general counterparts. To quote Iwasawa [Iw1]:

*"The theory of cyclotomic fields is in a unique position in algebraic number theory. On the one hand, it has provided us with a typical example of algebraic number fields from which we have been able to develop the theory of algebraic number fields in general; and on the other hand, it has also revealed to us many beautiful properties of the cyclotomic fields which are proper to these fields and which give us deep insights into important arithmetic results in elementary number theory."*

Already, it is known that the ideas discussed in this book work in some measure for elliptic curves over certain abelian $p$-adic Lie extensions, both for curves with complex multiplication ([CW2], [Ru], [Y]) and without complex multiplication ([Ka2], [SU]). It does not seem unreasonable to hope that this may turn out to be true in much greater generality, perhaps even in the direction of the non-abelian main conjecture made in [CFKSV].

Finally, we thank Karl Rubin for his very helpful comments on the manuscript.

# Contents

# Notation

## General Notations

The symbols $\mathbb{Z}$, $\mathbb{Q}$, $\mathbb{R}$ and $\mathbb{C}$ will denote the ring of integers, and the fields of rational numbers, real numbers and complex numbers, respectively. For a prime number $p$, we write $\mathbb{Q}_p$ for the completion of $\mathbb{Q}$ with respect to the $p$-adic valuation, normalized so that $|p|_p = p^{-1}$, and $\mathbb{C}_p$ for the completion of some fixed algebraic closure of $\mathbb{Q}_p$. As usual, $\mathbb{Z}_p$ will denote the ring of $p$-adic integers in $\mathbb{Q}_p$.

If $L/F$ is a Galois extension of fields, we write $\mathrm{Gal}(L/F)$ for the Galois group of $L$ over $F$.

If $m$ is an integer $\geq 1$, $\mu_m$ will denote the group of $m$-th roots of unity lying in some fixed algebraic closure of $\mathbb{Q}$ or $\mathbb{Q}_p$.

If $R$ is a ring, $R^\times$ will denote the multiplicative group of units of $R$. If $\Delta$ is a group, and $R$ is a commutative ring, we write $R[\Delta]$ for the group ring of $\Delta$ with coefficients in $R$.

## Specific Notations

It may help the reader to bear in mind the following notational convention. In general, we shall use script capital Latin symbols (e.g. $\mathcal{F}$, $\mathcal{K}$, $\mathcal{U}, \cdots$) to denote objects associated with the field which is generated over $\mathbb{Q}$ by all $p$-power roots of unity, while plain capital Latin symbols (e.g. $F$, $K$, $U, \cdots$) will denote the analogous objects attached to the corresponding maximal real subfield.

We now list some of the most commonly used symbols in the text. Let $n$ be either an integer $\geq 0$ or $\infty$.

$$\mathcal{F}_n = \mathbb{Q}(\mu_{p^{n+1}}), \quad F_n = \mathbb{Q}(\mu_{p^{n+1}})^+$$
$$\mathcal{K}_n = \mathbb{Q}_p(\mu_{p^{n+1}}), \quad K_n = \mathbb{Q}_p(\mu_{p^{n+1}})^+$$
$$\mathcal{G} = \mathrm{Gal}(\mathcal{F}_\infty/\mathbb{Q}), \quad G = \mathrm{Gal}(F_\infty/\mathbb{Q})$$

$\mathcal{L}_n =$ maximal abelian $p -$ extension of $\mathcal{F}_n$ unramified everywhere
$L_n =$ maximal abelian $p -$ extension of $F_n$ unramified everywhere
$\mathcal{M}_n =$ maximal abelian $p -$ extension of $\mathcal{F}_n$ unramified outside $p$
$M_n =$ maximal abelian $p -$ extension of $F_n$ unramified outside $p$

$$\mathcal{X}_\infty = \mathrm{Gal}(\mathcal{M}_\infty/\mathcal{F}_\infty),\ X_\infty = \mathrm{Gal}(M_\infty/F_\infty)$$
$$\mathcal{Y}_\infty = \mathrm{Gal}(\mathcal{L}_\infty/\mathcal{F}_\infty),\ \ Y_\infty = \mathrm{Gal}(L_\infty/F_\infty).$$

For integers $n$ with $0 \le n < \infty$, we define

$$\mathcal{U}_n = \text{group of units of } \mathcal{K}_n$$
$$U_n = \text{group of units of } K_n$$
$$D_n = \text{group of cyclotomic units of } F_n$$
$$V_n = \text{group of units of } F_n$$
$$C_n = \text{closure of } D_n \text{ in } U_n$$
$$E_n = \text{closure of } V_n \text{ in } U_n.$$

In the following definitions, the projective limits are taken with respect to the norm maps:-

$$\mathcal{U}_\infty = \varprojlim\ \mathcal{U}_n$$
$$U_\infty = \varprojlim\ U_n$$
$$C_\infty = \varprojlim\ C_n$$
$$E_\infty = \varprojlim\ E_n.$$

Finally, a superscript 1 on any of these objects indicates that it has been formed with the subgroup consisting of those elements which are congruent to 1 modulo the unique prime ideal above $p$.

# 1

# Cyclotomic Fields

## 1.1 Introduction

Let $p$ be an odd prime number. We owe to Kummer the remarkable discovery that there is a connexion between the arithmetic of the field generated over $\mathbb{Q}$ by the $p$-th roots of unity and the values of the Riemann zeta function at the odd negative integers. This arose out of his work on Fermat's last theorem. Almost a hundred years later, Iwasawa made the equally major discovery that the $p$-adic analogue of the Riemann zeta function is deeply intertwined with the arithmetic of the field generated over $\mathbb{Q}$ by all $p$-power roots of unity. The main conjecture, which is now a theorem (first completely proved by Mazur and Wiles [MW]), is the natural final outcome of these ideas. This main conjecture is the deepest result we know about the arithmetic of cyclotomic fields. In this first chapter, we explain more fully this background, and also give the precise statement of the main conjecture towards the end of the chapter. However, all proofs will be postponed until the later chapters.

Let $\mu_p$ denote the group of $p$-th roots of unity, and put

$$\mathcal{F} = \mathbb{Q}(\mu_p), \quad \varpi = \mathrm{Gal}(\mathcal{F}/\mathbb{Q}). \tag{1.1}$$

Now $\varpi$ acts on $\mu_p$, and thus gives an injective homomorphism

$$\theta : \varpi \hookrightarrow \mathrm{Aut}(\mu_p) = (\mathbb{Z}/p\mathbb{Z})^\times \tag{1.2}$$

In fact, $\theta$ is an isomorphism by the irreducibility of the $p$-th cyclotomic polynomial. Thus the powers $\theta^n$ for $n = 1, \ldots, p-1$ give all the characters of $\varpi$ with values in $\mathbb{F}_p$. Let $\mathfrak{C}$ denote the ideal class group of $\mathcal{F}$. We stress that $\mathfrak{C}$ becomes impossible to compute numerically by naive methods once $p$ is at all large. However, as is explained below, we owe to Kummer the discovery of a miraculous connexion between the $p$-primary subgroup of $\mathfrak{C}$ and the values

$$\zeta(s) \text{ for } s = -1, -3, -5, \ldots, \tag{1.3}$$

where $\zeta(s)$ is the classical complex Riemann zeta function. We recall that $\zeta(s)$ is defined by the Euler product

$$\zeta(s) = \prod_l (1 - l^{-s})^{-1} \tag{1.4}$$

for complex $s$ with real part greater than 1, and has an analytic continuation over the whole complex plane, apart from a simple pole at $s = 1$. It has been known since Euler that the values (1.3) are rational numbers. In fact,

$$\zeta(-n) = -\mathcal{B}_{n+1}/(n+1) \qquad (n = 1, 3, 5, \cdots) \tag{1.5}$$

where the Bernoulli numbers $\mathcal{B}_n$ are defined by the expansion

$$t/(e^t - 1) = \sum_{n=0}^{\infty} \mathcal{B}_n t^n /n!. \tag{1.6}$$

One computes easily from these equations that

$$\zeta(-1) = -\frac{1}{12}, \quad \zeta(-3) = \frac{1}{120}, \quad \zeta(-5) = -\frac{1}{252}, \cdots.$$

**Definition 1.1.1.** *We say that the prime number $p$ is irregular if $p$ divides the order of $\mathfrak{C}$.*

The first few irregular primes are $p = 37, 59, 67, 101, 103, \cdots$. It would be very difficult numerically to test whether a prime number $p$ is irregular if we did not have the following remarkable criterion for irregularity due to Kummer.

**Theorem 1.1.2.** *The prime $p$ is irregular if and only if $p$ divides the numerator of at least one of $\zeta(-1), \zeta(-3), \cdots, \zeta(4 - p)$.*

For example, we have

$$\zeta(-11) = \frac{691}{32760}, \quad \zeta(-15) = \frac{3617}{8160},$$

and thus, thanks to Kummer's theorem, we conclude that both 691 and 3617 are irregular primes. The irregularity of 37 follows from the fact that

$$\zeta(-31) = \frac{37 \times 208360028141}{16320}.$$

We point out that the numerators and denominators of these zeta values tend to grow very rapidly. For example, the numerator of $\zeta(-179)$ has 199 digits. However, fortunately Kummer's theorem basically reduces the problem of deciding whether a prime $p$ is irregular to a question of arithmetic modulo $p$, and is numerically very powerful. Indeed, using computational techniques derived from Kummer's theorem, all irregular primes up to 12,000,000 have been determined [BCEMS]. One finds that, up to this limit, the percentage of regular primes is approximately 60.61 percent, which fits remarkably well with the distribution which would occur if the numerators of the zeta values occurring in Theorem 1.1.2 were random modulo $p$ (see the discussion after Theorem 5.17 in [Wa]).

This mysterious link given in Theorem 1.1.2 between two totally different mathematical objects, namely the ideal class group of $\mathcal{F}$ on the one hand, and the special values of the Riemann zeta function on the other, is unquestionably one of the great discoveries in number theory, whose generalization to other arithmetic situations is a major theme of modern arithmetic geometry.

We end this introduction by recalling the following remarkable congruences, which were first discovered by Kummer as part of his proof of Theorem 1.1.2, and which provide the first evidence for the existence of the $p$-adic analogue of $\zeta(s)$.

**Theorem 1.1.3.** *Let $n$ and $m$ be odd positive integers such that $n \equiv m \not\equiv -1 \bmod (p-1)$. Then the rational numbers $\zeta(-n)$ and $\zeta(-m)$ are $p$-integral, and*

$$\zeta(-n) \equiv \zeta(-m) \bmod p.$$

## 1.2 Herbrand-Ribet Theorem

The beginning of the deeper understanding of Kummer's criterion for irregularity comes from considering the action of the Galois group $\varpi$ on $\mathfrak{C}$. After some work in this direction by both Kummer and Stickelberger, Herbrand considered the following specific refinement of Kummer's criterion. Let $\mathfrak{V} = \mathfrak{C}/\mathfrak{C}^p$, which is a finite dimensional vector space over the field $\mathbb{F}_p$, on which the Galois group $\varpi$ acts in a natural fashion. This action is semi-simple, because the order of $\varpi$ is prime to p. It is therefore natural to ask which of the characters $\theta^n$, where $n = 1, \ldots, p-1$, occur in $\mathfrak{V}$, and what is their multiplicity when they do occur? The theorem below, first established by Herbrand in one direction [He], and by Ribet [Ri] in the other, is today one of the important consequences of the main conjecture for the field $\mathcal{F}$.

**Theorem 1.2.1.** *Assume that $n$ is an odd integer with $3 \leq n \leq p - 2$. Then $\theta^n$ occurs in $\mathfrak{V} = \mathfrak{C}/\mathfrak{C}^p$ if and only if $p$ divides the numerator of $\zeta(n + 1 - p)$.*

Note that Theorem 1.2.1 says nothing about the occurrence in $\mathfrak{V}$ of $\theta^n$ for even integers $n$. In fact, no prime number $p$ has ever been found for which an even power of $\theta$ does occur in $\mathfrak{V}$, and Vandiver's conjecture asserts that no such $p$ exists. As we shall explain later, the main conjecture itself would be an easy consequence of a theorem of Iwasawa if Vandiver's conjecture were true. However, as far as we know, the main conjecture itself implies nothing in the direction of Vandiver's conjecture. Thus it is perhaps fair to say that Vandiver's conjecture seems inaccessible in our present state of knowledge, although it has been verified for all $p$ less than 12,000,000 in [BCEMS].

Here are some numerical examples of Theorem 1.2.1. For the irregular primes $p = 37, 59, 67, 101, 103, 131, 149$, $\mathfrak{V}$ has dimension 1 over $\mathbb{F}_p$. For the next irregular prime, namely $p = 157$, $\mathfrak{V}$ has dimension 2 over $\mathbb{F}_p$, with two distinct powers of $\theta$ occurring in it. A much more exotic example is given by $p = 12613$. In this example, $\mathfrak{V}$ has dimension 4 over $\mathbb{F}_p$, and 4 distinct powers of $\theta$ occur, namely

$$\theta^n, \text{ with } n \equiv 2077, \ 3213, \ 12111, \ 12305 \bmod 12612. \qquad (1.7)$$

In fact, the decomposition of $\mathfrak{V}$ into eigenspaces for the action of $\varpi$ is completely determined for all $p$ less than 12,000,000 in [BCEMS]. For such $p$, the characters which occur always have multiplicity 1, and the largest dimension of $\mathfrak{V}$ is 7.

## 1.3 The Cyclotomic Tower

Iwasawa's great insight was that one could go much further in explaining the above links by undertaking a seemingly more complicated study of the infinite tower of fields generated over $\mathbb{Q}$ by all $p$-power roots of unity. Although on the face of it, this will lead us to more elaborate and inaccessible arithmetic objects, the great benefit is that these objects are endowed with a natural action of the Galois group of the field generated over $\mathbb{Q}$ by all $p$-power roots of unity, which in the end can explain more easily and completely their relationship to the $p$-adic analogue of $\zeta(s)$.

Let $n$ be a natural number, and write $\mu_{p^{n+1}}$ (respectively, $\mu_{p^\infty}$) for the group of all $p^{n+1}$-th (resp. all $p$-power) roots of unity in some fixed algebraic closure of $\mathbb{Q}$. We define

$$\mathcal{F}_n = \mathbb{Q}(\mu_{p^{n+1}}), \quad \mathcal{F}_\infty = \mathbb{Q}(\mu_{p^\infty}), \tag{1.8}$$

and let

$$F_n = \mathbb{Q}(\mu_{p^{n+1}})^+, \quad F_\infty = \mathbb{Q}(\mu_{p^\infty})^+ \tag{1.9}$$

be their respective maximal totally real subfields (i.e. the fixed fields of the element induced by complex conjugation in their respective Galois groups over $\mathbb{Q}$). We write

$$\mathcal{G} = \mathrm{Gal}(\mathcal{F}_\infty/\mathbb{Q}), \quad G = \mathrm{Gal}(F_\infty/\mathbb{Q}) \tag{1.10}$$

for the corresponding Galois groups over $\mathbb{Q}$. The action of $\mathcal{G}$ on $\mu_{p^\infty}$ defines an injection

$$\chi : \mathcal{G} \longrightarrow \mathbb{Z}_p^\times = \mathrm{Aut}(\mu_{p^\infty}) \tag{1.11}$$

which is an isomorphism by the irreducibility of the cyclotomic equation. In particular, both $\mathcal{G}$ and $G$ are abelian. Let $\mathcal{L}_\infty$ (resp. $L_\infty$) be the maximal abelian $p$-extension of $\mathcal{F}_\infty$ (resp. $F_\infty$) which is unramified everywhere. Note that, since $p$ is always assumed to be odd, there is never any ramification of the primes at infinity in a $p$-extension. Put

$$\mathcal{Y}_\infty = \mathrm{Gal}(\mathcal{L}_\infty/\mathcal{F}_\infty), \quad Y_\infty = \mathrm{Gal}(L_\infty/F_\infty). \tag{1.12}$$

Since $\mathcal{Y}_\infty$ (resp. $Y_\infty$) is abelian, the Galois group $\mathcal{G}$ (resp. $G$) acts on it by inner automorphisms as follows. If $\sigma$ is an element of $\mathcal{G}$ (resp. $G$), pick any lifting $\tilde{\sigma}$ to the Galois group of $\mathcal{L}_\infty$ (resp. $L_\infty$) over $\mathbb{Q}$, and define $\sigma.y = \tilde{\sigma} y \tilde{\sigma}^{-1}$ for $y$ in $\mathcal{Y}_\infty$ (resp. $Y_\infty$). We remark that this is a very typical example of such a Galois action occurring in Iwasawa theory, and below we shall encounter another example of this kind.

The Iwasawa algebras of $\mathcal{G}$ and $G$ (see Appendix) are defined by

$$\Lambda(\mathcal{G}) = \varprojlim \mathbb{Z}_p[\mathcal{G}/\mathcal{H}], \quad \Lambda(G) = \varprojlim \mathbb{Z}_p[G/H],$$

where $\mathcal{H}$ (resp. $H$) runs over the open subgroups of $\mathcal{G}$ (resp. $G$). Since $\mathcal{Y}_\infty$ (resp. $Y_\infty$) is by construction a compact $\mathbb{Z}_p$-module, the $\mathcal{G}$-action (resp. $G$-action) on it extends by continuity and linearity to an action of the whole Iwasawa algebra $\Lambda(\mathcal{G})$ (resp. $\Lambda(G)$) (see Appendix). Standard arguments in Iwasawa theory show that $\mathcal{Y}_\infty$ (resp. $Y_\infty$) is a finitely generated torsion module over $\Lambda(\mathcal{G})$ (resp. $\Lambda(G)$).

We digress briefly here to point out that the two Iwasawa modules $\mathcal{Y}_\infty$ and $Y_\infty$ have a very different nature arithmetically. In fact, $Y_\infty = 0$ if Vandiver's conjecture is true for $p$ (and hence, in particular, for all $p < 12,000,000$). However, $\mathcal{Y}_\infty$ has positive $\mathbb{Z}_p$-rank precisely when the prime $p$ is irregular. In addition, an important theorem of Ferrero-Washington [Fe-W] shows that both $\mathcal{Y}_\infty$ and $Y_\infty$ are always finitely generated $\mathbb{Z}_p$-modules. Let $\mathcal{J} = \{1, \iota\}$ be the subgroup of $\mathcal{G}$ fixing $F_\infty$. Since $p$ is odd, there is a decomposition

$$\mathcal{Y}_\infty = \mathcal{Y}_\infty^+ \oplus \mathcal{Y}_\infty^- \tag{1.13}$$

as $\Lambda(\mathcal{G})$-modules, where the complex conjugation $\iota$ acts on the first direct summand by $+1$ and on the second by $-1$. In fact, it is easily seen that the natural surjection from $\mathcal{Y}_\infty$ onto $Y_\infty$ induces an isomorphism from $\mathcal{Y}_\infty^+$ onto $Y_\infty$. Even after taking this decomposition, the discrepancies between these two modules continue. For example, it is known that $\mathcal{Y}_\infty^-$ is a free finitely generated $\mathbb{Z}_p$-module. On the other hand, it is an important unsolved problem about the tower of fields $\mathcal{F}_\infty$, whether or not $Y_\infty$ has any non-zero finite submodule which is stable under the action of $G$. In fact, as we shall see in Chapters 4 and 6, the maximal finite $G$-submodule of $Y_\infty$ plays an important role in the completion of the proof of the main conjecture using Euler systems. One of the beauties of the main conjecture is that it can be proven for all $p$, irrespective of knowing the answers to these finer questions.

## 1.4 The Main Conjecture

The main conjecture could in fact be stated in terms of the $\Lambda(\mathcal{G})$-module $\mathcal{Y}_\infty^-$ of the previous section. However, because of the method of proof that we shall follow, it is more natural to work with an equivalent version in terms of a different Iwasawa module. For this reason, we consider larger abelian extensions of the fields $\mathcal{F}_\infty$ and $F_\infty$. Let $\mathcal{M}_\infty$ (resp. $M_\infty$) be the maximal abelian $p$-extension of $\mathcal{F}_\infty$ (resp. $F_\infty$) which is unramified outside the unique prime above $p$ in $\mathcal{F}_\infty$ (resp. $F_\infty$). We write

$$\mathcal{X}_\infty = \mathrm{Gal}(\mathcal{M}_\infty/\mathcal{F}_\infty), \quad X_\infty = \mathrm{Gal}(M_\infty/F_\infty). \tag{1.14}$$

In an entirely similar manner to that described earlier, $\mathcal{G}$ (resp. $G$) acts on $\mathcal{X}_\infty$ (resp. $X_\infty$) via inner automorphisms, making it a module over $\Lambda(\mathcal{G})$ (resp. over $\Lambda(G)$). While both these modules are finitely generated over the respective Iwasawa algebras, the module $\mathcal{X}_\infty$ is not $\Lambda(\mathcal{G})$-torsion whereas $X_\infty$ is $\Lambda(G)$-torsion by the following important theorem due to Iwasawa [Iw4].

**Theorem 1.4.1.** *The module $X_\infty$ is a finitely generated torsion $\Lambda(G)$-module.*

Before stating the Main Conjecture, whose formulation requires some results from the structure theory of $\Lambda(G)$-modules, it is perhaps interesting to again digress and note (although neither fact is needed for the version given below), that the theorem of Ferrero-Washington [Fe-W] implies that $X_\infty$ is a finitely generated $\mathbb{Z}_p$-module, and that a theorem of Iwasawa [Iw4] implies that $X_\infty$ has no non-zero $\mathbb{Z}_p$-torsion (see Proposition 4.7.2). Thus $X_\infty$ is a free finitely generated $\mathbb{Z}_p$-module, on which the group $G$, which is topologically generated by one element, is acting continuously. One could then take the characteristic polynomial of some topological generator of $G$ acting on $X_\infty$ as a generator of the characteristic ideal of $X_\infty$. However, we shall work without assuming these stronger results, and simply recall that the structure theory of finitely generated torsion $\Lambda(G)$-modules (see Appendix) implies that for each such module $N$, there is an exact sequence of $\Lambda(G)$-modules

$$0 \longrightarrow \bigoplus_{i=1}^{r} \frac{\Lambda(G)}{\Lambda(G)f_i} \longrightarrow N \longrightarrow D \longrightarrow 0,$$

where $f_i$ $(i = 1, \cdots, r)$ is a non-zero divisor, and $D$ is finite. Then the characteristic ideal of $N$, which we denote by $\mathrm{ch}_G(N)$, is defined to be the ideal of $\Lambda(G)$ generated by the product $f_1 \ldots f_r$.

It is at first utterly surprising that, as we now explain, there is a generator of $\mathrm{ch}_G(X_\infty)$ which is intimately related to the Riemann zeta function. We owe to Kubota-Leopoldt [KL] the first proof that a $p$-adic analogue of $\zeta(s)$ exists. Iwasawa then discovered [Iw3] that this $p$-adic analogue, which we denote by $\zeta_p$, has a natural interpretation in terms of the Iwasawa algebra $\Lambda(G)$. As will be shown in Chapter 3, the elements of $\Lambda(G)$ can be viewed as $\mathbb{Z}_p$-valued measures on the Galois group $G$. To take account of the fact that $\zeta_p$ has, like the Riemann zeta function, a simple pole, one defines a pseudo-measure on $G$ to be any element $\mu$ of the ring of fractions of $\Lambda(G)$ such that $(g - 1)\mu$ belongs to $\Lambda(G)$ for all $g$ in $G$ (see §3.2). The integral

$$\int_G \nu\, d\mu$$

of any non-trivial continuous homomorphism $\nu : G \longrightarrow \mathbb{Z}_p^\times$ against a pseudo-measure $\mu$ is then well-defined.

**Theorem 1.4.2.** *There exists a unique pseudo-measure $\zeta_p$ on $G$ such that*

$$\int_G \chi(g)^k d\zeta_p = (1 - p^{k-1})\zeta(1 - k)$$

*for all even integers $k \geq 2$.*

As hinted at above, $\zeta_p$ has a simple pole at the trivial character, with residue $1 - p^{-1}$, in the following sense. Write $\kappa$ for the composition of the cyclotomic character $\chi$ with the natural projection from $\mathbb{Z}_p^\times$ to the multiplicative group $1 + p\mathbb{Z}_p$. It is clear that $\kappa$ factors through $G$, and that it makes sense to raise it to any power in $\mathbb{Z}_p$. Then it can be shown that, if $s$ is an element of $\mathbb{Z}_p$ distinct from 1, we have an expansion of the form

$$\int_G \kappa^{1-s} d\zeta_p = (1 - p^{-1})(s-1)^{-1} + a_0 + a_1(s-1) + \cdots,$$

where $a_0, a_1, \cdots$ are elements of $\mathbb{Z}_p$. This is closely related to the classical von-Staudt-Clausen theorem on Bernoulli numbers.

Let $I(G)$ denote the kernel of the augmentation homomorphism from $\Lambda(G)$ to $\mathbb{Z}_p$. As $\zeta_p$ is a pseudo-measure, $I(G)\zeta_p$ is an ideal of $\Lambda(G)$.

**Theorem 1.4.3.** (Main Conjecture) *We have*

$$\mathrm{ch}_G(X_\infty) = I(G)\zeta_p.$$

The first complete proof was given by Mazur-Wiles [MW] using the arithmetic of modular curves. A second proof, based on the generalization of Ribet's proof of Theorem 1.2.1, was given by Wiles [W]. The goal of this book is to give what is probably the simplest proof of this theorem, which proceeds along the following lines. We first establish Iwasawa's theorem (see the next section) for a $\Lambda(G)$-module closely related to $X_\infty$, and then use arguments from Euler systems due to Kolyvagin, Rubin and Thaine [Ko], [Ru3], [Th], to show that the discrepancy between these two modules does not alter their characteristic ideals. In fact, this discrepancy is zero for all known numerical examples, including all $p < 12,000,000$.

## 1.5 Iwasawa's Theorem

The genesis of the main conjecture is Iwasawa's paper [Iw2], and his important theorem below arises from combining the results of this paper with his construction of $\zeta_p$ in [Iw3]. For each $n \geq 0$, consider now the local field

$$K_n = \mathbb{Q}_p(\mu_{p^{n+1}})^+. \tag{1.15}$$

We write $U_n^1$ for the group of units of $K_n$, which are $\equiv 1 \bmod \mathfrak{p}_n$, where $\mathfrak{p}_n$ is the maximal ideal of the ring of integers of $K_n$. Let $D_n$ be the group of cyclotomic units of $F_n$. Thus $D_n$ is generated by all Galois conjugates of

$$\pm \frac{\zeta_n^{-e/2} - \zeta_n^{e/2}}{\zeta_n^{-1/2} - \zeta_n^{1/2}}$$

where $\zeta_n$ denotes a primitive $p^{n+1}$-th root of unity, and $e$ is a primitive root modulo $p$ such that $e^{p-1} \not\equiv 1 \bmod p^2$. We define $D_n^1$ to be the subgroup of all elements of $D_n$ which are $\equiv 1 \bmod \mathfrak{p}_n$. Finally, let

$$C_n^1 = \overline{D}_n^1$$

be the closure of $D_n^1$ in $U_n^1$ with respect to the $\mathfrak{p}_n$-adic topology. Define

$$U_\infty^1 = \varprojlim U_n^1, \quad C_\infty^1 = \varprojlim C_n^1, \tag{1.16}$$

where the projective limits are taken with respect to the norm maps. Of course, the group $G$ acts continuously on both these $\mathbb{Z}_p$-modules, endowing them with an action of $\Lambda(G)$. Iwasawa's theorem is the following:-

**Theorem 1.5.1.** *The $\Lambda(G)$-module $U_\infty^1/C_\infty^1$ is canonically isomorphic to $\Lambda(G)/I(G) \cdot \boldsymbol{\zeta}_p$, where $\boldsymbol{\zeta}_p$ is the p-adic zeta function, and $I(G)$ is the augmentation ideal.*

We shall give a very elementary proof of this theorem, different from Iwasawa's (see Theorem 4.4.1), which does not even use local class field theory. This proof was discovered by Wiles and one of us [CW2] when studying the analogous theorem for elliptic curves with complex multiplication. However, we follow Coleman's beautiful proof [Co] of the existence of the interpolating power series lying behind this approach, rather than using the ad hoc method of [CW2].

The comparison between the Galois group $X_\infty$ and the module $U_\infty^1/C_\infty^1$ is provided by class field theory. Let $V_n^1$ be the group of units of the ring of integers of $F_n$ which are $\equiv 1 \bmod \mathfrak{p}_n$, and define

$$E_n^1 = \overline{V}_n^1, \quad E_\infty^1 = \varprojlim E_n^1 \tag{1.17}$$

where the closure in $U_n^1$ is again taken with respect to the $p$-adic topology and the projective limit is taken with respect to the norm maps. As we explain in more detail in § 4.5, the Artin map of global class field theory gives a canonical $\Lambda(G)$-isomorphism

$$\mathrm{Gal}(M_\infty/L_\infty) \simeq U_\infty^1/E_\infty^1.$$

Thus we have the four term exact sequence of $\Lambda(G)$-modules

$$0 \longrightarrow E_\infty^1/C_\infty^1 \longrightarrow U_\infty^1/C_\infty^1 \longrightarrow X_\infty \longrightarrow Y_\infty \longrightarrow 0, \qquad (1.18)$$

all of which are finitely generated torsion modules. But the characteristic ideal is multiplicative in exact sequences (see Appendix). Hence, granted Iwasawa's theorem, we have the following result.

**Proposition 1.5.2.** *The main conjecture is true if and only if* $\mathrm{ch}_G(Y_\infty) = \mathrm{ch}_G(E_\infty^1/C_\infty^1)$.

Of course, this last proposition does not involve the $p$-adic zeta function $\zeta_p$, and is only of real interest when combined with Iwasawa's theorem via the exact sequence (1.18). The proof of Proposition 1.5.2 using Euler systems is given in Chapters 4 and 5, and broadly follows Rubin's generalization of the method discovered by Kolyvagin and Thaine. It is striking that this proof largely uses ideas already known to Kummer, combined with global class field theory.

We end this chapter by making some brief remarks about applications of the main conjecture. However, we omit detailed proofs in this book because these applications are dealt with rather fully in the literature, and also because some of them involve higher $K$-theory.

We first explain why Kummer's criterion for irregularity is a consequence of the main conjecture. Using an important result of Iwasawa (see Proposition 4.7.2), which asserts that $X_\infty$ has no non-zero finite $\Lambda(G)$-submodule, it follows easily from the main conjecture and the structure theory of finitely generated torsion $\Lambda(G)$-modules (see Appendix), that

$$I(G)\zeta_p = \Lambda(G) \qquad (1.19)$$

if and only if

$$X_\infty = 0. \qquad (1.20)$$

However, we claim that (1.19) is equivalent to the assertion that all of the values

$$\zeta(-1), \ \zeta(-3), \cdots, \zeta(4-p) \qquad (1.21)$$

are $p$-adic units. Indeed, as is explained in the Appendix (see (A2)), for any finitely generated torsion $\Lambda(G)$-module $M$, we have a decomposition

$$M = \bigoplus_{i \bmod \frac{p-1}{2}} M^{(i)},$$

where $M^{(i)}$ denotes the submodule of $M$ on which $G(F_0/\mathbb{Q})$ acts via $\theta^{2i}$. As was mentioned earlier (see the discussion after Theorem 1.4.2), $\zeta_p$ has a simple pole with residue $1 - p^{-1}$ at the trivial character, from which it follows easily that

$$(I(G)\zeta_p)^{(0)} = \Lambda(G)^{(0)}.$$

On the other hand, taking $i$ to be any of $1, \cdots, (p-3)/2$, and combining Theorem 1.4.2 with Lemma 3.6.2, we see that

$$(I(G)\zeta_p)^{(i)} = \Lambda(G)^{(i)}$$

if and only if $p$ does not divide the numerator of $\zeta(1-2i)$. In particular, it follows that (1.19) is valid if and only if all the values in (1.21) are $p$-adic units.

Next, we must relate (1.20) to the ideal class group of the field $\mathcal{F}_0 = \mathbb{Q}(\mu_p)$. For each $n \geq 0$, let $\mathcal{A}_n$ denote the $p$-primary part of the ideal class group of $\mathcal{F}_n$ and define

$$\mathcal{A}_\infty = \varinjlim \mathcal{A}_n,$$

where the inductive limit is taken with respect to the natural maps coming from the inclusion of fields. As always, we write $\mathcal{A}_\infty^-$ for the submodule of $\mathcal{A}_\infty$ on which complex conjugation acts by $-1$. To relate $X_\infty$ to $\mathcal{A}_\infty^-$, we invoke the following isomorphism coming from multiplicative Kummer theory (see, for example, [C1]). There is a canonical $\mathcal{G}$-isomorphism

$$\mathcal{A}_\infty^- = \mathrm{Hom}(X_\infty, \mu_{p^\infty}). \tag{1.22}$$

Moreover, it is known that the natural map from $\mathcal{A}_n^-$ to $\mathcal{A}_\infty^-$ is injective and induces an isomorphism

$$\mathcal{A}_n^- \simeq (\mathcal{A}_\infty^-)^{\Gamma_n}$$

for all $n \geq 0$, where $\Gamma_n = \mathrm{Gal}(\mathcal{F}_\infty/\mathcal{F}_n)$ [Iw1]. But, for any discrete $p$-primary $\Gamma_0$-module $N$, $N^{\Gamma_0} = 0$ if and only if $N = 0$. In view of these remarks, we see that

$$\mathcal{A}_0^- \neq 0 \text{ if and only if } X_\infty \neq 0. \tag{1.23}$$

To complete the proof of Kummer's criterion given in Theorem 1.2.1, one has to prove the stronger statement that $\mathcal{A}_0 \neq 0$ if and only if

$X_\infty \neq 0$. One direction is proved by (1.23). Conversely, assume that $\mathcal{A}_0^+ \neq 0$, or equivalently that $p$ divides the class number of $F_0$. Thus, writing $L_0$ for the $p$-Hilbert class field of $F_0$, we have $L_0 \neq F_0$, and so $L_0 F_\infty \neq F_\infty$ because $F_\infty/F_0$ is totally ramified at the unique prime above $p$. But clearly, $L_0 F_\infty$ is contained in $M_\infty$, and so $X_\infty \neq 0$ as required. A slight refinement of this argument, in which one considers eigenspaces for the action of the subgroup $\varpi$ of $\mathcal{G}$ of order $p-1$ on these modules, enables one to prove the Herbrand-Ribet Theorem 1.2.1.

Finally, the applications to $K$-theory arise from the fact that the higher $K$-groups of the rings of integers of finite extensions of $\mathbb{Q}$ contained in $F_\infty$ can be related to the module $\mathcal{A}_\infty^-$ twisted by positive powers of the cyclotomic character $\chi$ of $\mathcal{G}$.

# 2

# Local Units

## 2.1 Introduction

The aim of this chapter will be to study various aspects of the local units at the prime above $p$ of the cyclotomic tower $\mathcal{F}_\infty = \mathbb{Q}(\mu_{p^\infty})$, with a view to preparing the ground for the proof of Iwasawa's theorem (Theorem 4.4.1) in Chapter 4. In fact, these local results are interesting in their own right, and have connexions with the theory of Fontaine and the $K$-theory of group rings, see for example [F], [O], [BM]. Our basic tool will be the construction of canonical interpolation series for norm compatible systems of elements in the tower $\mathcal{F}_\infty$. These interpolation series were discovered in the course of the work [CW] by Wiles and one of us, the original proof there being rather ad hoc. Almost immediately, Coleman [Co] found a beautiful conceptual proof, which is valid for arbitrary Lubin-Tate groups. We will give Coleman's proof for the formal multiplicative group in this chapter. We will also discuss the $p$-adic logarithmic derivatives of these norm compatible systems, which are also defined in [CW] but have their antecedents in the work of Kummer and Takagi (see [H] for a brief account), and which are important for the construction of the $p$-adic zeta function.

Let $n$ be a natural number, and define $\mathcal{K}_n = \mathbb{Q}_p(\mu_{p^{n+1}})$, where we recall that $\mu_{p^{n+1}}$ denotes the group of $p^{n+1}$-th roots of unity. Write $\mathcal{U}_n$ for the multiplicative group of units of the ring of integers of $\mathcal{K}_n$. We fix for the rest of this chapter a generator $\zeta_n$ of $\mu_{p^{n+1}}$, with the property that $\zeta_{n+1}^p = \zeta_n$ for all $n \geq 0$, and put

$$\pi_n = \zeta_n - 1.$$

Thus $(\zeta_n)$ is a generator of the free $\mathbb{Z}_p$-module of rank 1 defined by

$$T_p(\mu) = \varprojlim \mu_{p^{n+1}},$$

where the transition maps are given by raising to the $p$-th power. Moreover, $\pi_n$ is a local parameter for $\mathcal{K}_n$. Given any $z$ in $\mathcal{U}_n$, there is therefore a power series $f(T)$ in the ring $R = \mathbb{Z}_p[[T]]$ of formal power series in the variable $T$ with coefficients in $\mathbb{Z}_p$ such that $f(\pi_n) = z$. From the very early days of the theory of cyclotomic fields, mathematicians tried to exploit this fact to define a derivative of $z$ by differentiating the formal power series $f(T)$ with respect to $T$. The problem is that $f(T)$ is in no way uniquely determined by $z$, and one only obtains a weak notion of the derivative because of this lack of uniqueness. It was realized in [CW] that this difficulty could be overcome by considering all $n$ simultaneously.

Denote the norm map from the multiplicative group of $\mathcal{K}_n$ to $\mathcal{K}_m$ by $\mathrm{N}_{n,m}$ when $n \geq m$; plainly $\mathrm{N}_{n,n-1}$ maps $\mathcal{U}_n$ into $\mathcal{U}_{n-1}$.

**Definition 2.1.1.** We define $\mathcal{U}_\infty = \varprojlim \mathcal{U}_n$, where the projective limit is taken with respect to the norm maps.

The goal of this chapter is to prove the following theorem.

**Theorem 2.1.2.** *For each* $\mathbf{u} = (u_n)$ *in* $\mathcal{U}_\infty$, *there exists a unique* $\mathrm{f}_{\mathbf{u}}(T)$ *in* $R$ *such that* $\mathrm{f}_{\mathbf{u}}(\pi_n) = u_n$ *for all* $n \geq 0$.

The following classical example originally suggested the above general theorem. Let $a$ and $b$ be non-zero integers which are relatively prime to $p$, and define

$$\mathbf{u} = (u_n), \quad \text{where } u_n = \frac{\zeta_n^{-a/2} - \zeta_n^{a/2}}{\zeta_n^{-b/2} - \zeta_n^{b/2}}.$$

It is an easy classical exercise to see that $u_n$ is a unit in $\mathcal{F}_n$ such that $\mathrm{N}_{n,m}(u_n) = u_m$ for all $n \geq m$. Moreover, the power series

$$w_k(T) = \frac{(1+T)^{-k/2} - (1+T)^{k/2}}{T}$$

is a unit in $R$ whenever $(k, p) = 1$. Hence the power series

$$\mathrm{f}_{\mathbf{u}}(T) = w_a(T)/w_b(T)$$

belongs to $R$, and satisfies $\mathrm{f}_{\mathbf{u}}(\pi_n) = u_n$, proving in this special case the existence of the power series as in the above theorem.

In fact, the uniqueness of the power series $\mathrm{f}_{\mathbf{u}}(T)$ in Theorem 2.1.2 is immediate from the Weierstrass preparation theorem [Bou, Chapter VII], which we now recall. A polynomial $g(T)$ in $\mathbb{Z}_p[T]$ is defined to be distinguished if it is monic and all its lower coefficients belong to $p\mathbb{Z}_p$.

**Theorem 2.1.3.** *Each $f(T)$ in $R$ can be written uniquely in the form $f(T) = p^m g(T) w(T)$, where $m$ is a non-negative integer, $g(T)$ is a distinguished polynomial, and $w(T)$ is a unit in $R$.*

Note that a power series $f(T)$ in $R$ converges on the maximal ideal of the ring of integers of the algebraic closure of $\mathbb{Q}_p$. As a unit in $R$ clearly has no zeroes, it follows from the Weierstrass preparation theorem that each power series in $R$ has only a finite number of zeroes in the maximal ideal of the ring of integers of the algebraic closure of $\mathbb{Q}_p$.

## 2.2 Norm and Trace Operators

In his proof of Theorem 2.1.2, Coleman introduced some interesting new operators on the ring $R$, which we now explain. We remark that these operators have subsequently been vastly generalized in Fontaine's theory of $(\varphi, \Gamma)$-modules [F]. From now on, the ring $R$ is endowed with the topology defined by the powers of the maximal ideal $\mathfrak{m} = (p, T)$.

**Definition 2.2.1.** For $f$ in $R$, we define $\varphi(f)(T) = f((1+T)^p - 1)$.

Clearly $\varphi$ is a $\mathbb{Z}_p$-algebra endomorphism of $R$.

**Lemma 2.2.2.** *The map $\varphi$ is injective.*

*Proof.* We need only remark that if $h(T) = a_n T^n + \cdots$ with $a_n \neq 0$, then $\varphi(h)(T) = p^n a_n T^n + \cdots$, which is clearly non-zero.  $\square$

We note that if $\xi$ belongs to $\mu_p$ and $f(T) = \sum_{n=0}^{\infty} a_n T^n$, it is easily seen that $f(\xi(1+T) - 1) = \sum_{n=0}^{\infty} a_n (\xi(1+T) - 1)^n$ converges to an element of $\mathcal{O}[[T]]$, where $\mathcal{O}$ is the ring of integers of $\mathbb{Q}_p(\mu_p)$. Recall that $R^\times$ is the group of units of $R$.

**Proposition 2.2.3.** *There exist unique continuous maps*

$$\mathcal{N} \,:\, R \longrightarrow R, \quad \psi \,:\, R \longrightarrow R$$

*such that*

$$(\varphi \circ \mathcal{N})(f)(T) = \prod_{\xi \in \mu_p} f(\xi(1+T) - 1), \tag{2.1}$$

$$(\varphi \circ \psi)(f)(T) = \frac{1}{p} \cdot \sum_{\xi \in \mu_p} f(\xi(1+T) - 1). \tag{2.2}$$

*Moreover, $\psi$ is a $\mathbb{Z}_p$-module homomorphism, $\psi \circ \varphi = 1_R$, and $\mathcal{N}$ preserves products. In particular, $\mathcal{N}$ maps $R^\times$ to itself.*

In fact, the uniqueness of $\mathcal{N}$ and $\psi$ is plain from the injectivity of $\varphi$. To prove the existence, we shall need the following lemma.

**Lemma 2.2.4.** *The image of $\varphi$ consists of all power series $h(T)$ in $R$ satisfying*

$$h(\xi(1+T)-1) = h(T) \text{ for all } \xi \text{ in } \mu_p. \tag{2.3}$$

*Proof.* It is clear that every power series in $\varphi(R)$ satisfies (2.3). Conversely, let $h(T)$ be any element of $R$ satisfying (2.3). Since $h(\xi-1) - h(0) = 0$ for all $\xi$ in $\mu_p$, the Weierstrass preparation theorem shows that

$$h(T) - h(0) = \varphi(T)h_1(T)$$

for some $h_1(T)$ in $R$. Let $n$ be any integer $\geq 1$. Assume that we have already found $a_0, \cdots, a_{n-1}$ in $\mathbb{Z}_p$ such that

$$h(T) = \sum_{i=0}^{n-1} a_i\varphi(T)^i + \varphi(T)^n h_n(T) \tag{2.4}$$

with $h_n(T)$ in $R$. Clearly, we again have that $h_n(\xi(1+T)-1) = h_n(T)$, whence, applying the earlier observation with $h(T)$ replaced by $h_n(T)$, it follows that assertion (2.4) also holds for $n+1$. By induction, the proof of the lemma is complete.     □

Let us first use this lemma to prove the existence of the operator $\mathcal{N}$. Given $f$ in $R$, define $h(T) = \prod_{\xi \in \mu_p} f(\xi(1+T)-1)$, which is clearly also in $R$. Clearly $h(T) = h(\xi(1+T)-1)$ for all $\xi$ in $\mu_p$, and so by the above lemma $h(T) = \varphi(g(T))$ for some $g(T)$ in $R$. Thus we can take $\mathcal{N}(f) = g$. The existence of the operator $\psi$ is a little more complicated because of the factor $1/p$. We define

$$r(T) = \sum_{\xi \in \mu_p} f(\xi(1+T)-1).$$

It is clear that $r(T)$ belongs to $R$, and we must show that

$$r(T) = p \cdot s(T)$$

for some $s(T)$ in $R$. Let $\mathfrak{p}_0$ be the maximal ideal of the ring of integers of $\mathbb{Q}_p(\mu_p)$. Since for each $\xi$ in $\mu_p$, we have

$$\xi(1+T)-1 \equiv T \mod \mathfrak{p}_0 R,$$

it follows that $r(T)$ must indeed belong to $pR$, as claimed above. Again, it is then clear that $s(\xi(1+T)-1) = s(T)$ for all $\xi$ in $\mu_p$ and so

$s(T) = \varphi(q(T))$ for some $q(T)$ in $R$. We can therefore define $\psi(f) = q$. It is clear that $\psi \circ \varphi = 1_R$. The final assertions of the proposition are then clear and hence the proof of the proposition is complete. $\qquad\square$

**Lemma 2.2.5.** *We have* $\mathcal{N}(T) = T$. *For all integers* $n \geq 1$, *we have*

$$\psi\left(\varphi(T)^n \cdot \left(\frac{1+T}{T}\right)\right) = T^n \cdot \left(\frac{1+T}{T}\right).$$

*Proof.* The first assertion is clear from the injectivity of $\varphi$ and the identity

$$\varphi(T) = \prod_{\xi \in \mu_p} (\xi(1+T) - 1).$$

Taking the logarithmic derivative with respect to $T$ of both sides of this identity, and then multiplying both sides by $(1+T)\varphi(T)^n$, we obtain the new identity

$$\varphi\left(T^n \cdot \left(\frac{1+T}{T}\right)\right) = \frac{1}{p} \sum_{\xi \in \mu_p} h(\xi(1+T) - 1),$$

where

$$h(T) = \frac{1+T}{T} \cdot \varphi(T)^n.$$

Since $\varphi$ is injective, the second assertion of the lemma is now clear from Proposition 2.2.3. $\qquad\square$

## 2.3 Interpolating Power Series

The aim of this section is to give Coleman's proof of Theorem 2.1.2. The essential idea is to look for units $f$ in $R$ with $\mathcal{N}(f) = f$. Indeed, if $f$ is in $R^\times$, then it is plain that $f(\pi_n)$ belongs to $\mathcal{U}_n$ for all $n \geq 0$. In addition, as $\mathcal{N}(f) = f$, these values are norm compatible, for the following reason. The minimal equation of $\zeta_n$ over $\mathcal{K}_{n-1}$ is $X^p - \zeta_{n-1} = 0$, and thus

$$N_{n,n-1} f(\zeta_n - 1) = \prod_{\xi \in \mu_p} f(\xi\zeta_n - 1). \tag{2.5}$$

On the other hand, the equation $\mathcal{N}(f) = f$ can be rewritten, by virtue of (2.1) as

$$\varphi(f)(T) = \prod_{\xi \in \mu_p} f(\xi(1+T) - 1).$$

Since, by definition, $\varphi(f)(\pi_n) = f(\pi_{n-1})$, it follows from the above displayed formula that the right hand side of (2.5) is indeed equal to

$f(\pi_{n-1})$. We now turn to the harder part of the proof, which is to show that all elements of $\mathcal{U}_\infty$ are obtained in this manner.

The natural way to find fixed points of the operator $\mathcal{N}$ acting on the group $R^\times$ of units of $R$ is to show that, for any unit $f$ in $R$, the limit of the sequence of $\mathcal{N}^k(f)$ always exists in $R$ as $k$ tends to infinity. This will be shown now via a sequence of lemmas.

**Lemma 2.3.1.** *Assume $f$ is in $R$ and let $k \geq 0$ be an integer. If $\varphi(f)(T) \equiv 1 \bmod p^k R$, then $f(T) \equiv 1 \bmod p^k R$.*

*Proof.* Write

$$f(T) - 1 = \left( \sum_{n=0}^{\infty} a_n T^n \right) p^m$$

where not all of the $a_n$ are divisible by $p$ and $m \geq 0$ is an integer. Let $r$ be the smallest integer such that $p \nmid a_r$. We have

$$\varphi(f)(T) - 1 = p^m h(T), \text{ where } h(T) = \sum_{n=0}^{\infty} a_n \varphi(T)^n.$$

Now $\varphi(T) \equiv T^p \bmod pR$, and so we have

$$h(T) \equiv a_r T^{pr} + \cdots \bmod pR.$$

Hence, as $p \nmid a_r$, $h(T)$ is not in $pR$, and so our hypothesis implies that $m \geq k$. $\qquad\square$

**Lemma 2.3.2.** *Assume $f \in R^\times$. Then $\mathcal{N}(f) \equiv f \bmod pR$. If we assume further that $f \equiv 1 \bmod p^m R$ for some integer $m \geq 1$, then $\mathcal{N}(f) \equiv 1 \bmod p^{m+1} R$.*

*Proof.* Let $\mathfrak{p}_0$ be the maximal ideal of the ring of integers of $\mathbb{Q}_p(\mu_p)$. Suppose that $f \equiv 1 \bmod p^k R$, for some integer $k \geq 0$. In other words, if $f(T) = \sum_{n=0}^{\infty} a_n T^n$, we have $a_0 \equiv 1 \bmod p^k$ and $a_n \equiv 0 \bmod p^k$ for $n \geq 1$. Since for each $\xi$ in $\mu_p$, we have

$$\xi(1 + T) - 1 \equiv T \bmod \mathfrak{p}_0 R,$$

it follows that

$$f(\xi(1 + T) - 1) \equiv f(T) \bmod \mathfrak{p}_0 p^k R.$$

Thus

$$\varphi(\mathcal{N}(f)) = \prod_{\xi \in \mu_p} f(\xi(1+T)-1) \equiv f(T)^p \bmod p^{k+1}R. \qquad (2.6)$$

Assume first that $k \geq 1$. Then plainly, $f(T)^p \equiv 1 \bmod p^{k+1}R$ and the assertion of the lemma follows from Lemma 2.3.1. If $k = 0$, we note that

$$f(T)^p \equiv f(T^p) \equiv \varphi(f)(T) \quad \bmod pR.$$

Thus again the conclusion of the lemma follows from Lemma 2.3.1.    □

**Corollary 2.3.3.** *Assume $f$ is in $R^\times$, and let $k_2 \geq k_1 \geq 0$. Then $\mathcal{N}^{k_2}(f) \equiv \mathcal{N}^{k_1}(f) \bmod p^{k_1+1}R$.*

*Proof.* To establish the Corollary, note that $\mathcal{N}^{k_2-k_1}(f)/f \equiv 1 \bmod pR$ by Lemma 2.3.2. Applying $\mathcal{N}^{k_1}$ to both sides, the corollary follows from the second assertion of the same lemma.    □

**Corollary 2.3.4.** *If $f$ is any element of $R^\times$, then $g = \lim_{k \to \infty} \mathcal{N}^k(f)$ exists in $R^\times$ and $\mathcal{N}(g) = g$.*

*Proof.* The ring $R$ is complete in the topology defined by the powers of the maximal ideal $\mathfrak{m} = (p, T)$, whence the assertion is clear from the previous corollary.    □

Now we can at last prove Theorem 2.1.2. Let $\mathbf{u}$ be any element of $\mathcal{U}_\infty$. For each $n \geq 0$, choose $f_n$ in $R^\times$ such that $f_n(\pi_n) = u_n$, and consider the sequence $\{g_n\}$ in $R$ where $g_n(T) = \mathcal{N}^n f_{2n}(T)$. Since $R$ is compact with respect to its topology, this sequence has at least one limit point which we denote by $h(T)$. The following lemma shows that $h(T)$ satisfies $h(\pi_n) = u_n$ for all $n \geq 0$, whence $h(T)$ can be taken to be $f_\mathbf{u}(T)$, thereby completing the proof of Theorem 2.1.2.

**Lemma 2.3.5.** *For all $n \geq 0$, and all $m \geq n$, we have $g_m(\pi_n) \equiv u_n \bmod p^{m+1}$. In particular, $\lim_{m \to \infty} g_m(\pi_n) = u_n$.*

*Proof.* Since $u_{n-1} = N_{n,n-1}(u_n)$, we conclude that $u_{n-1} = (\mathcal{N}f_n)(\pi_{n-1})$. Repeating this $k$ times for $1 \leq k \leq n$, we find

$$u_{n-k} = N_{n,n-k}(f_n(\pi_n)) = (\mathcal{N}^k f_n)(\pi_{n-k}).$$

Suppose now that $m \geq n$. We obtain

$$u_n = (\mathcal{N}^{2m-n} f_{2m})(\pi_n).$$

But by Corollary 2.3.3, we have

$$\mathcal{N}^{2m-n} f_{2m} \equiv \mathcal{N}^m f_{2m} \bmod p^{m+1} R.$$

Evaluating both sides of this congruence at $\pi_n$, we therefore conclude that $u_n \equiv g_m(\pi_n) \bmod p^{m+1} R.$ □

We end this section by introducing an action of $\mathcal{G} = \mathrm{Gal}(\mathcal{K}_\infty/\mathbb{Q}_p)$ where $\mathcal{K}_\infty = \mathbb{Q}_p(\mu_{p^\infty})$, on the ring $R$. Recall that

$$\chi \; : \; \mathcal{G} \longrightarrow \mathbb{Z}_p^\times$$

is the cyclotomic character defined by $\sigma(\zeta) = \zeta^{\chi(\sigma)}$ for all $\sigma \in \mathcal{G}$ and $\zeta$ in $\mu_{p^\infty}$. For $\sigma$ in $\mathcal{G}$, define

$$(\sigma f)(T) = f((1+T)^{\chi(\sigma)} - 1) \quad (f \in R). \tag{2.7}$$

This gives a group action of $\mathcal{G}$ on the ring $R$ which maps $R^\times$ to itself. Further, it is clear that $\varphi(\sigma f) = \sigma(\varphi f)$ for all $\sigma$ in $\mathcal{G}$ and $f$ in $R$, and that this action of $\mathcal{G}$ commutes with the operator $\mathcal{N}$ and $\psi$.

**Definition 2.3.6.** *We define $W$ to be the set of all elements $f$ in $R^\times$ such that $\mathcal{N}(f) = f$.*

The following corollary is immediate from our proof of Theorem 2.1.2.

**Corollary 2.3.7.** *The map $\mathbf{u} \mapsto f_{\mathbf{u}}(T)$ defines a $\mathcal{G}$-isomorphism from $\mathcal{U}_\infty$ onto $W$.*

## 2.4 The Logarithmic Derivative

The aim of this section is to study some delicate properties of logarithmic differentiation, which relates the multiplicative unit group $R^\times$ to the additive group of $R$.

**Lemma 2.4.1.** *We have $(1 - \varphi)R = TR$.*

*Proof.* The inclusion of $(1 - \varphi)R$ in $TR$ is plain. Conversely, if $h$ is any element of $TR$, let us show that it lies in $(1 - \varphi)R$. For each $n \geq 0$, we define $\omega_n(T) = (1+T)^{p^n} - 1$. As this is a distinguished polynomial of degree $p^n$, the division lemma part of the Weierstrass preparation theorem [Bou] shows that we can write

$$h = h_n + \omega_n r_n,$$

where $h_n$ is a polynomial in $\mathbb{Z}_p[T]$ of degree less than $p^n$, and $r_n$ is an element of $R$. Define

$$l_n = \sum_{i=0}^{n-1} \varphi^i(h_{n-i}).$$

Clearly, we have

$$l_{n+1} - \varphi(l_n) = h_{n+1}.$$

Since $h_{n+1}$ converges to $h$ in $R$, it suffices to show that $l_n$ converges to some $l$ in $R$, because then we would have $h = (1 - \varphi)l$. Now, for $k = 1, \cdots, n$, we have

$$\varphi^{n-k}(h) = \varphi^{n-k}(h_k) + \omega_n \varphi^{n-k}(r_k).$$

Adding these equations for $k = 1, \cdots, n$, we obtain the identity

$$\sum_{i=0}^{n} \varphi^i(h) = l_n + \omega_n s_n,$$

for some $s_n$ in $R$. As $h$ is in $TR$, it is clear that the sum on the left hand side converges, as $n$ tends to infinity. This completes the proof of the lemma.    $\square$

**Definition 2.4.2.** *We define the following subsets of $R$:-*

$$R^{\psi=1} = \{f \in R \ : \ \psi(f) = f\}; \quad R^{\psi=0} = \{f \in R \ : \ \psi(f) = 0\}.$$

**Lemma 2.4.3.** *There exists an exact sequence*

$$0 \longrightarrow \mathbb{Z}_p \longrightarrow R^{\psi=1} \overset{\theta}{\longrightarrow} R^{\psi=0} \longrightarrow \mathbb{Z}_p \longrightarrow 0, \qquad (2.8)$$

*where $\theta(f) = (1 - \varphi)(f)$, and where the map on the left is the natural inclusion, while the map on the right is evaluation at $T = 0$.*

*Proof.* Note first that $\theta$ maps $R^{\psi=1}$ to $R^{\psi=0}$ because $\psi \circ \varphi$ is the identity map on $R$. It is also clear that the image of $\mathbb{Z}_p$ is contained in the kernel of $\theta$, and that the image of $\theta$ is contained in the kernel of the map on the right. Also, the map on the right is surjective, since, for example, $1 + T$ belongs to $R^{\psi=0}$ (as follows easily, for example, from Proposition 2.2.3). By Lemma 2.4.1, the ideal $TR$ is the image of $\theta$. Hence we need only show exactness at $R^{\psi=1}$, and, as remarked earlier, $\mathbb{Z}_p$ lies in the kernel of $\theta$. If $f(T)$ is not in $\mathbb{Z}_p$, it will be of the form

$$f(T) = b_0 + b_r T^r + \cdots, \text{ where } b_r \neq 0.$$

But then

$$\varphi(f(T)) = b_0 + p^r b_r T^r + \cdots,$$

and so clearly $\varphi(f) \neq f$, and the proof of the lemma is complete.    □

Recall that $W$ denotes the subset of $R^\times$ consisting of all units $f$ such that $\mathcal{N}(f) = f$. We now discuss the relationship between $W$ and the subset $R^{\psi=1}$ of the additive group of $R$. Denote the formal derivative with respect to $T$ of any $f(T)$ in $R$ by $f'(T)$.

**Definition 2.4.4.** *For $f$ in $R^\times$, define*

$$\Delta(f) = (1+T)\frac{f'(T)}{f(T)}.$$

It is clear that $\Delta$ is a group homomorphism from $R^\times$ to the additive group of $R$.

**Lemma 2.4.5.** *We have $\Delta(W) \subset R^{\psi=1}$. Further, the kernel of $\Delta$ on $W$ is the group $\mu_{p-1}$ of the $(p-1)$-th roots of unity.*

*Proof.* Let $f$ be in $W$. Recalling that $\varphi(f)(T) = f((1+T)^p - 1)$ and applying $\Delta$ to the equation

$$\varphi(f) = \prod_{\xi \in \mu_p} f(\xi(1+T)-1),$$

we obtain immediately that $\psi(\Delta(f)) = \Delta(f)$. The final assertion of the lemma is obvious.    □

In fact, the following stronger result is true, but its proof is subtle and non-trivial.

**Theorem 2.4.6.** *We have $\Delta(W) = R^{\psi=1}$.*

The strategy of the proof is to use reduction modulo $p$. Let

$$\Omega = R/pR = \mathbb{F}_p[[T]], \tag{2.9}$$

and let $x \mapsto \tilde{x}$ be the reduction map. If $Y$ is any subset of $R$, then we denote by $\widetilde{Y}$ its image in $\Omega$ under the reduction map.

**Lemma 2.4.7.** *If $\widetilde{\Delta(W)} = \widetilde{R^{\psi=1}}$, then $\Delta(W) = R^{\psi=1}$.*

*Proof.* Assume that the reductions of $\Delta(W)$ and $R^{\psi=1}$ do coincide, and take any $g$ in $R^{\psi=1}$. Hence there exists $h_1$ in $W$ such that $\widetilde{\Delta(h_1)} = \tilde{g}$. This implies that $\Delta(h_1) - g = pg_2$ for some $g_2$ in $R$, and again we have that $\psi(g_2) = g_2$. Repeating this argument, we conclude that there exists $h_2$ in $W$ such that $\Delta(h_2) - g_2 = pg_3$, with $g_3$ in $W$. Note that since $\Delta(a) = 0$ for all $a$ in $\mathbb{Z}_p^\times$, it can be assumed, by multiplying by an appropriate $(p-1)$-th root of unity, that $h_1$, $h_2$, $\cdots$, all have constant term which is congruent to 1 modulo $p$. But clearly,

$$\Delta\left(\frac{h_1}{h_2^p}\right) = g - p^2 g_3,$$

and, continuing in this manner, we get a sequence of elements $h_1$, $h_1/h_2^p, \cdots$. of $W$ which converges to $h$ in $W$ with $\Delta(h) = g$.  □

**Lemma 2.4.8.** *We have $\widetilde{W} = \Omega^\times$.*

*Proof.* If $x$ is any element of $\Omega^\times$, we can find $f$ in $R^\times$ such that $\tilde{f} = x$. But by Corollary 2.3.4, $g = \lim_{k \to \infty} \mathcal{N}^k(f)$ exists in $R^\times$, and $\mathcal{N}(g) = g$. On the other hand, by Corollary 2.3.3, $\mathcal{N}^k(f) \equiv f \mod pR$ for all $k \geq 1$, and hence $\tilde{g} = \tilde{f}$, thereby proving the lemma.  □

The delicate part of the proof of Theorem 2.4.6 is to determine $\widetilde{R^{\psi=1}}$. To this end, consider the map

$$\partial \; : \; \Omega^\times \longrightarrow \Omega$$

defined by $\partial(g) = T \cdot g'(T)/g(T)$.

**Lemma 2.4.9.** *We have $\widetilde{R^{\psi=1}} = \left(\frac{1+T}{T}\right) \partial(\Omega^\times)$.*

Let us first remark that Theorem 2.4.6 plainly follows from the above three lemmas on noting that for $f$ in $R^\times$, we have $\widetilde{\Delta(f)} = (1+T)/T \cdot \partial \tilde{f}$.

We begin the proof of Lemma 2.4.9. We first need the following result:-

**Lemma 2.4.10.** *We have $\partial(\Omega^\times) = \Phi$, where $\Phi = \{f = \sum_{n=1}^\infty a_n T^n :$ $a_n = a_{np}\}$ for all $n \geq 1$.*

Let $\Theta$ be the subset of $\Omega$ consisting of all series of the form $f = \sum_{n=1}^\infty a_n T^n$ with $a_n = 0$ for all $n$ with $(n,p) = 1$.

**Corollary 2.4.11.** *We have $T\Omega = \partial(\Omega^\times) + \Theta$.*

To deduce the corollary, take any power series $g = \sum_{n=1}^{\infty} b_n T^n$ in $T\Omega$, and define

$$h = \sum_{\substack{m=1 \\ (m,p)=1}}^{\infty} b_m \sum_{k=0}^{\infty} T^{mp^k}.$$

It is plain that $g - h$ belongs to $\Theta$, and by the above lemma, $h$ is in $\Phi$.

*Proof of Lemma 2.4.10.* Let us first note that every element $f(T)$ in $\Omega^\times$ can be written as a convergent infinite product

$$f(T) = a \prod_{n=1}^{\infty} (1 - a_n T^n)$$

where $a$ is non-zero, and all $a_n$ are in $\mathbb{F}_p$. Recall that this is proven by the usual inductive argument as follows. We can assume that $a = 1$, and that $f(T)$ is of the form

$$f(T) = 1 + c_r T^r + \cdots .$$

Then one sees that

$$f(T)(1 + c_r T^r)^{-1} = 1 + d_{r+1} T^{r+1} + \cdots ,$$

for some $d_{r+1}$ in $\mathbb{F}_p$, and we continue in this manner. In view of this infinite product expansion, to prove that $\partial(\Omega^\times)$ is contained in $\Phi$, it suffices to show that $\partial(1 - a_k T^k)$ is contained in $\Phi$ for all $k \geq 1$. But this is easily seen to be true from the explicit formula

$$\partial(1 - a_k T^k) = -k \sum_{m=1}^{\infty} a_k^m T^{mk}.$$

Conversely, suppose that $h = \sum_{n=1}^{\infty} d_n T^n$ is any element of $\Phi$. We claim that, for each integer $m \geq 1$, there exist elements $e_1, \cdots, e_{m-1}$ in $\mathbb{F}_p$ such that

$$h_m = h - \partial(1 - e_1 T) - \cdots - \partial(1 - e_{m-1} T^{m-1})$$

belongs to $T^m \Omega$. We prove this by induction, it being trivially true for $m = 1$. Assume it is true for $m$, and write

$$h_m = d_m T^m + \cdots .$$

If $d_m = 0$, we simply take $e_m = 0$. Suppose now that $d_m \neq 0$. As $h_m$ belongs to $\Phi$ (because $h$ does), it follows that necessarily $(m, p) = 1$. Thus we can solve in $\mathbb{F}_p$ the linear equation

$$m e_m = -d_m.$$

Since

$$\partial(1 - e_m T^m) = d_m T^m + \cdots,$$

our inductive hypothesis for $m + 1$ has been proven. Hence, defining

$$g = \prod_{n=1}^{\infty} (1 - e_n T^n),$$

we have $\partial(g) = h$. This completes the proof of Lemma 2.4.10.    □

*Proof of Lemma 2.4.9.* As $\Delta(W) \subset R^{\psi=1}$, it follows from Lemma 2.4.8 that $((1 + T)/T) \cdot \partial(\Omega^{\times})$ is contained in $\widetilde{R^{\psi=1}}$. Conversely, take any $f$ in $R^{\psi=1}$ and let

$$g = \tilde{f}, \quad s = \left(\frac{T}{T+1}\right) g.$$

Thus we must show that $s$ belongs to $\partial(\Omega^{\times})$. Now $s$ belongs to $T\Omega$, and so by the above Corollary, we have

$$s = \partial(w) + h$$

where $w$ is in $\Omega^{\times}$ and $h = \sum_{m=1}^{\infty} d_m T^{mp}$ is in $\Theta$. Rewrite this equation as

$$g = \left(\frac{T+1}{T}\right) \partial(w) + k$$

where

$$k = \sum_{m=1}^{\infty} d_m \left(\frac{T+1}{T}\right) T^{mp}.$$

Since $\psi$ is $\mathbb{Z}_p$-linear, it induces a map $\tilde{\psi} : \Omega \longrightarrow \Omega$. Now $\tilde{\psi}$ fixes $g$ by hypothesis and $\tilde{\psi}$ fixes $((T + 1)/T) \partial(w)$ by Lemma 2.4.8. Hence by the above equation $\tilde{\psi}(k) = k$. On the other hand, for all $n \geq 1$,

$$\widetilde{\varphi(T)}^n = T^{pn}.$$

Therefore Lemma 2.2.5 shows that

$$\tilde{\psi}\left(T^{pn}\left(\frac{1+T}{T}\right)\right) = T^n\left(\frac{1+T}{T}\right).$$

Since $\widetilde{\psi}(k) = k$, it follows that $h = \sum_{m=1}^{\infty} d_m T^m$. But, if $h$ is non-zero, this clearly contradicts the fact that $h$ lies in $\Theta$. Indeed, if $d_m$ is non-zero for some integer $\geq 1$, then, writing $m = p^r m'$, where $r \geq 0$, and $m'$ is primes to $p$, it follows that $d_{m'} = d_m$ is also non-zero, which is impossible because $h$ lies in $\Theta$. Hence h = 0, and thus the proof of the lemma, and so also Theorem 2.4.6, is complete. $\qquad\square$

## 2.5 An Exact Sequence

Our goal in this section is to study a canonical map, which was first introduced in [CW2], but which has its origin in Leopoldt's theory of the $\Gamma$-transform [L]. This canonical map will be the key to proving Iwasawa's theorem.

We define straightaway the canonical map in question.

**Lemma 2.5.1.** *For all $f$ in $R^{\times}$, the series*

$$\mathcal{L}(f) := \frac{1}{p} \log \left( \frac{f(T)^p}{\varphi(f)(T)} \right) \qquad (2.10)$$

*lies in $R$. If $f$ lies in $W$, then $\mathcal{L}(f)$ lies in $R^{\psi=0}$. The map $\mathcal{L} : W \longrightarrow R^{\psi=0}$ thus defined is a $\mathcal{G}$-homomorphism with $\mathcal{G}$ acting on $W$ and on $R^{\psi=0}$ by (2.7).*

*Proof.* For any $f$ in $R^{\times}$, we clearly have

$$\varphi(f) \equiv f(T)^p \bmod pR.$$

Hence, writing $g(T) = \frac{f(T)^p}{\varphi(f)(T)}$, it follows that

$$g(T) = 1 + ph(T)$$

for some $h(T)$ in $R$. Now $p^{n-1}/n$ lies in $\mathbb{Z}_p$ for all $n = 1, \cdots$, and thus

$$\log(g(T)) = \sum_{n=1}^{\infty} \frac{(-1)^{n-1} p^n h(T)^n}{n}$$

converges to an element of $pR$, proving that $\mathcal{L}(f)$ lies in $R$. It is clear that $\mathcal{L}$ is a $\mathcal{G}$-homomorphism.

We now show that $\mathcal{L}(f)$ lies in $R^{\psi=0}$ when $f$ is in $W$. Since every element of $R$ can be written as a product of an element in $\mu_{p-1}$ and a power series whose constant term is congruent to 1 modulo $p$, we may clearly assume that the constant term of $f$ is congruent to 1 modulo $p$.

Hence the series $\log(f(T))$ is a well-defined element of $\mathbb{Q}_p[[T]]$. Since $f$ is in $W$, we have the equation

$$\varphi(f(T)) = \prod_{\xi \in \mu_p} f(\xi(1 + T) - 1).$$

Taking logarithms of both sides of this equation we deduce that

$$\log \varphi(f)(T) = \sum_{\xi \in \mu_p} \log f(\xi(1 + T) - 1).$$

Hence we obtain that

$$\sum_{\xi \in \mu_p} \mathcal{L}(f)(\xi(1 + T) - 1) = 0,$$

which shows by (2.2) that $\mathcal{L}(f)$ does indeed belong to $R^{\psi=0}$. This completes the proof of the lemma. □

Let $A$ be the subset of $R^\times$ defined by

$$A = \{\xi(1 + T)^a \ : \ \xi \in \mu_{p-1}, \ a \in \mathbb{Z}_p\}.$$

Let $D$ be the differential operator on $R$ given by $D(f) = (1 + T)f'(T)$.

**Theorem 2.5.2.** *There is a canonical exact sequence of $\mathcal{G}$-modules*

$$0 \longrightarrow A \longrightarrow W \xrightarrow{\mathcal{L}} R^{\psi=0} \xrightarrow{\alpha} \mathbb{Z}_p \longrightarrow 0 \qquad (2.11)$$

*where $\alpha$ is given by $\alpha(f) = (Df)(0)$.*

*Proof.* It is clear that $A \subset \ker(\mathcal{L})$. To prove the converse, let us note that if $f(T)$ is an element of $R$ with $f(0) \equiv 1 \bmod p$ and $\log f(T) = 0$, then $f(T) = 1$. Indeed, we can write $f(T) = bg(T)$ with $b \equiv 1 \bmod p$ and $g(T)$ of the form

$$g(T) = 1 + c_r T^r + \cdots$$

where $r \geq 1$ and $c_r \neq 0$. Hence

$$\log g(T) = c_r T^r + \cdots.$$

But $\log f(T) = 0$ gives

$$0 = \log b + \log g(T),$$

whence it is easily seen that $b = 1$ and $\log g(T) = 0$, which contradicts our hypothesis that $c_r \neq 0$.

Suppose now that $f(T)$ is any element of ker $\mathcal{L}$. Multiplying it by a suitable element of $\mu_{p-1}$, we may suppose that $f(0) \equiv 1 \bmod p$, and thus the same is true for $h(T) = f(T)^p/\varphi(f(T))$. But then, the above remark shows that $\mathcal{L}(f) = 0$ yields $h(T) = 1$. By Corollary 2.3.7, there exists a unique $\mathbf{u} = (u_n)$ in $\mathcal{U}_\infty$ such that $f = f_{\mathbf{u}}$ and hence we have

$$f_{\mathbf{u}}((1+T)^p - 1) = f_{\mathbf{u}}(T)^p.$$

This implies that $u_n^p = u_{n-1}$ for all $n \geq 1$ and that $f_{\mathbf{u}}(0)$ is in $\mu_{p-1}$. But then $f_{\mathbf{u}}(0) = 1$ since $f_{\mathbf{u}} \equiv 1 \bmod p$ and so $(u_n) \in T_p(\mu)$. Thus there exists $a$ in $\mathbb{Z}_p$ such that $\mathbf{u} = (\zeta_n)^a$, whence $f(T) = (1+T)^a$. This proves that $\ker(\mathcal{L}) = A$.

It is clear that $\alpha \circ \mathcal{L} = 0$, and the surjectivity of $\alpha$ follows from noting that $\psi(1+T) = 0$ and that $\alpha(1+T) = 1$. Hence it only remains to prove that $\ker(\alpha) \subset \mathrm{Im}(\mathcal{L})$, which is the delicate part of the proof of the theorem. We have the commutative diagram

$$
\begin{array}{ccc}
W & \xrightarrow{\ \mathcal{L}\ } & R^{\psi=0} \\
{\scriptstyle \Delta}\downarrow & & \downarrow{\scriptstyle D} \\
R^{\psi=1} & \xrightarrow{\ \theta\ } & R^{\psi=0},
\end{array}
$$

recalling that $\theta(f) = (1 - \varphi)(f)$.

Note that $D$ is clearly injective on $R^{\psi=0}$. Suppose $f$ is any element of $R^{\psi=0}$ with $\alpha(f) = 0$. Define $g = Df$ so that $g$ is in $TR$ by the definition of $\alpha$. Then Lemma 2.4.1 shows that there exists $h$ in $R^{\psi=1}$ with $\theta(h) = g$. We now invoke the key fact that $\Delta$ is surjective (cf. Theorem 2.4.6), to conclude that there exists $w$ in $W$ with $\Delta(w) = h$. By construction and the commutativity of the diagram, we have

$$g = D(f) = D\mathcal{L}(w).$$

Hence $f = \mathcal{L}(w)$ by the injectivity of $D$. Thus $f$ belongs to the image of $\mathcal{L}$ and the proof of the theorem is complete. $\qquad\square$

## 2.6 The Higher Logarithmic Derivative Maps

In this section, we use Theorem 2.1.2 to define the higher logarithmic derivatives of elements of $\mathcal{U}_\infty$. We study these maps and show by a mysterious, but elementary, calculation going back to Kummer, that the values of the Riemann zeta function at the odd negative integers arise as the higher logarithmic derivatives of cyclotomic units.

**Definition 2.6.1.** *For each integer $k \geq 1$, define the logarithmic derivative homomorphism $\delta_k : \mathcal{U}_\infty \longrightarrow \mathbb{Z}_p$ by*

$$\delta_k(\mathbf{u}) = \left( D^{k-1} \left( \frac{(1+T)\mathfrak{f}'_\mathbf{u}(T)}{\mathfrak{f}_\mathbf{u}(T)} \right) \right)_{T=0} \tag{2.12}$$

*where $\mathbf{u}$ is any element of $\mathcal{U}_\infty$, $\mathfrak{f}_\mathbf{u}(T)$ is the associated power series in Theorem 2.1.2, and the subscript $T = 0$ means evaluation at 0.*

We remark that $\delta_k$ takes values in $\mathbb{Z}_p$ because $\mathfrak{f}_\mathbf{u}$ is a unit in $R$. Also, recall that the Galois group $\mathcal{G}$ of $\mathcal{K}_\infty$ acts on $\mathcal{U}_\infty$ in the natural fashion, and on $R$ by (2.7).

**Lemma 2.6.2.** *For all $k \geq 1$, the map $\delta_k$ is a group homomorphism satisfying*

$$\delta_k(\sigma(\mathbf{u})) = \chi(\sigma)^k \delta_k(\mathbf{u})$$

*for all $\mathbf{u}$ in $\mathcal{U}_\infty$, and all $\sigma$ in $\mathcal{G}$.*

*Proof.* The first assertion is plain, and the second follows from the observation that

$$f_{\sigma(\mathbf{u})}(T) = \mathfrak{f}_\mathbf{u}((1+T)^{\chi(\sigma)} - 1),$$

and the following elementary identity

$$D^k(g((1+T)^a - 1)) = a^k (D^k g)((1+T)^a - 1), \tag{2.13}$$

for all $k \geq 0$, $a$ in $\mathbb{Z}_p$, and $g$ in $R$.    □

We now carry out the crucial calculation of $\delta_k$ on cyclotomic units, stressing that it is via this calculation that the values of the Riemann-zeta function at the odd negative integers appear first in our approach to the main conjecture. Let $a$ and $b$ be integers which are prime to $p$ and define $\mathbf{c}(a,b) = (c_n(a,b))$ by

$$c_n(a,b) = \frac{\zeta_n^{-a/2} - \zeta_n^{a/2}}{\zeta_n^{-b/2} - \zeta_n^{b/2}}. \tag{2.14}$$

We have already remarked that $\mathbf{c}(a,b)$ belong to $\mathcal{U}_\infty$. Let $\zeta(s)$ denote the complex Riemann zeta function.

**Proposition 2.6.3.** *We have*
*(i) $\delta_k(\mathbf{c}(a,b)) = 0$ for $k = 1, 3, 5 \cdots$,*
*(ii) $\delta_k(\mathbf{c}(a,b)) = (b^k - a^k)\zeta(1-k)$ for $k = 2, 4, 6, \cdots$.*

*Proof.* Put

$$f(T) = \frac{(1+T)^{-a/2} - (1+T)^{a/2}}{(1+T)^{-b/2} - (1+T)^{b/2}}$$

so that $f(\pi_n) = c_n(a,b)$ for all $n \geq 0$. We make the change of variable $T = e^z - 1$ so that $D = \frac{\mathrm{d}}{\mathrm{d}z}$. Hence we have

$$\delta_k(\mathbf{c}(a,b)) = \left( \left( \frac{\mathrm{d}}{\mathrm{d}z} \right)^{k-1} g(z) \right)_{z=0}$$

where

$$g(z) = \frac{\mathrm{d}}{\mathrm{d}z} \log f(e^z - 1).$$

But

$$2g(z) = b \left( \frac{1}{e^{-bz} - 1} - \frac{1}{e^{bz} - 1} \right) - a \left( \frac{1}{e^{-az} - 1} - \frac{1}{e^{az} - 1} \right)$$

By definition, we have

$$\frac{1}{e^t - 1} = \sum_{n=0}^{\infty} \frac{\mathcal{B}_n}{n!} \cdot t^{n-1},$$

where $\mathcal{B}_n$ is the $n$-th Bernoulli number. Hence, as $\mathcal{B}_m = 0$ when $m$ is odd with $m > 1$, we obtain easily that

$$g(z) = \sum_{\substack{k=2 \\ k \text{ even}}}^{\infty} \frac{\mathcal{B}_k z^{k-1}}{k!} (a^k - b^k).$$

But now the proposition follows from the well-known fact that

$$\zeta(1-k) = -\frac{\mathcal{B}_k}{k} \quad (k = 2,4,6 \cdots).$$

This completes the proof of the proposition.     $\square$

We end this chapter by proving a result about the image of the $\delta_k$ for $k = 1, \cdots, p-1$, which is closely related to the original proof of Theorem 2.1.2 given in [CW]. We refer the reader to Theorem 3.6.1 of the next chapter for a determination of the image of the $\delta_k$ for all $k \geq 1$, which curiously does not seem easy to prove by the methods given here.

**Theorem 2.6.4.** *For $k = 1, \cdots, p-1$, we have $\delta_k(\mathcal{U}_\infty) = \mathbb{Z}_p$.*

*Proof.* Since, by the first lemma in this section, the image of $\delta_k$ is an ideal in $\mathbb{Z}_p$, it suffices to prove that there exists $\mathbf{u}$ in $\mathcal{U}_\infty$ such that $\delta_k(\mathbf{u})$ is a unit in $\mathbb{Z}_p^\times$ for $k = 1, \cdots, p-1$. We can clearly view $D$ as an operator on $\Omega$, and thus we must find a $\mathbf{u}$ in $\mathcal{U}_\infty$ such that

$$D^{k-1}(\widetilde{\Delta f_{\mathbf{u}}(T)})$$

has non-zero constant term for $k = 1, \cdots, p-1$. By the proof of Theorem 2.4.6 (see Lemmas 2.4.8, 2.4.9, 2.4.10), there exists $\mathbf{u}$ in $\mathcal{U}_\infty$ such that

$$\widetilde{\Delta(f_{\mathbf{u}}(T))} = (1+T)\alpha_1(T), \text{ where } \alpha_1(T) = \sum_{m=0}^\infty T^{p^m-1}.$$

We proceed to show that this $\mathbf{u}$ has the desired properties. Clearly, the series $(1+T)\alpha_1(T)$ has a non-zero constant term, proving the assertion for $k = 1$. For $k > 1$, define

$$\alpha_k(T) = \sum_{m=1}^\infty T^{p^m-k}.$$

Recalling that these series lie in $\Omega$, one verifies immediately that

$$\alpha_k'(T) = -k\alpha_{k+1}(T) \quad (k = 1, \cdots, p-1).$$

It follows that

$$D((1+T)^k\alpha_k(T)) = k((1+T)^k\alpha_k(T) - (1+T)^{k+1}\alpha_{k+1}(T)),$$

whence a simple inductive argument shows that

$$D^{k-1}((1+T)\alpha_1(T)) = (1+T)\alpha_1(T) + \sum_{j=2}^k c_j(1+T)^j\alpha_j(T)$$

for $k = 1, \cdots, p-1$, where the $c_j$ are elements of $\mathbb{F}_p$. The proof now follows on noting that $\alpha_j(0) = 0$ for $j = 2, \cdots, p-1$. $\qquad\square$

# 3

# Iwasawa Algebras and $p$-adic Measures

## 3.1 Introduction

In this chapter, we interpret the results on power series proven in the previous chapter in terms of $p$-adic measures on the Galois group $\mathcal{G} = \mathrm{Gal}(\mathbb{Q}(\mu_{p^\infty})/\mathbb{Q})$. We remark that the language of $p$-adic measures was first introduced in the paper [MSD]. A key tool in our reinterpretation will be Mahler's well known theorem on continuous $p$-adic functions on $\mathbb{Z}_p$. This leads us to an entirely equivalent reformulation of the canonical map $\mathcal{L}$ appearing in Theorem 2.5.2, and also the exact sequence there. As a consequence, we shall obtain in the next chapter both a simple construction of the $p$-adic analogue of the complex Riemann zeta function, and a proof of Iwasawa's Theorem 4.4.1.

The Iwasawa algebra of both $\mathcal{G}$ and the additive group of $\mathbb{Z}_p$ will play a fundamental role in all that follows, and we briefly recall their definition now. More generally, let $\mathfrak{G}$ be any profinite abelian group, and let $\mathcal{T}_{\mathfrak{G}}$ be the set of open subgroups of $\mathfrak{G}$. We define the Iwasawa algebra $\Lambda(\mathfrak{G})$ by

$$\varprojlim \mathbb{Z}_p[\mathfrak{G}/\mathfrak{H}]$$

where $\mathfrak{H}$ runs over $\mathcal{T}_{\mathfrak{G}}$, and $\mathbb{Z}_p[\mathfrak{G}/\mathfrak{H}]$ denotes the ordinary group ring over $\mathbb{Z}_p$. It is a compact topological $\mathbb{Z}_p$-algebra, the topology being the natural one on the projective limit coming from the $p$-adic topology on the group rings of the finite quotients of $\mathfrak{G}$.

## 3.2 $p$-adic Measures

As above, let $\mathfrak{G}$ be any profinite abelian group. We now sketch the proof that the elements of the Iwasawa algebra $\Lambda(\mathfrak{G})$ define integral $p$-adic measures on $\mathfrak{G}$.

Let $\mathbb{C}_p$ be the completion of the algebraic closure of the field of $p$-adic numbers $\mathbb{Q}_p$, and write $|\ |_p$ for its $p$-adic valuation. Let $C(\mathfrak{G}, \mathbb{C}_p)$ be the $\mathbb{C}_p$-algebra of all continuous functions from $\mathfrak{G}$ to $\mathbb{C}_p$. As usual, we can define a norm on $C(\mathfrak{G}, \mathbb{C}_p)$ by

$$\|f\| = \sup_{g \in \mathfrak{G}} |f(g)|_p,$$

and this makes $C(\mathfrak{G}, \mathbb{C}_p)$ into a $\mathbb{C}_p$-Banach space. We recall that a function $f$ in $C(\mathfrak{G}, \mathbb{C}_p)$ is defined to be locally constant if there exists an open subgroup $\mathfrak{H}$ of $\mathfrak{G}$ such that $f$ is constant modulo $\mathfrak{H}$, i.e. gives a function from $\mathfrak{G}/\mathfrak{H}$ to $\mathbb{C}_p$. Write $\mathrm{Step}(\mathfrak{G})$ for the sub-algebra of locally constant functions, which is easily seen to be everywhere dense.

We now explain how to integrate any continuous $\mathbb{C}_p$-valued function on $\mathfrak{G}$ against an element $\lambda$ of $\Lambda(\mathfrak{G})$. We begin with locally constant functions. Suppose that $f$ in $\mathrm{Step}(\mathfrak{G})$ is locally constant modulo the subgroup $\mathfrak{H}$ of $\mathfrak{G}$. Write $\lambda_{\mathfrak{H}}$ for the image of $\lambda$ in $\mathbb{Z}_p[\mathfrak{G}/\mathfrak{H}]$, say

$$\lambda_{\mathfrak{H}} = \sum_{x \in \mathfrak{G}/\mathfrak{H}} c_{\mathfrak{H}}(x) x, \tag{3.1}$$

where the $c_{\mathfrak{H}}(x)$ lie in $\mathbb{Z}_p$. We then define

$$\int_{\mathfrak{G}} f d\lambda = \sum_{x \in \mathfrak{G}/\mathfrak{H}} c_{\mathfrak{H}}(x) f(x).$$

One sees easily that this is independent of the choice of $\mathfrak{H}$. Since the $c_{\mathfrak{H}}(x)$ lie in $\mathbb{Z}_p$, we have

$$\left| \int_{\mathfrak{G}} f d\lambda \right|_p \leq \|f\|. \tag{3.2}$$

We further note that, if $\varepsilon_x$ denotes the characteristic function of the coset $x$ in $\mathfrak{G}/\mathfrak{H}$, then we have

$$\int_{\mathfrak{G}} \varepsilon_x d\lambda = c_{\mathfrak{H}}(x). \tag{3.3}$$

Let $f$ be any continuous $\mathbb{C}_p$-valued function $f$ on $\mathfrak{G}$. We can choose a sequence $\{f_n\}$ in $\mathrm{Step}(\mathfrak{G})$ which converges to $f$. It is plain from (3.2) that the sequence of integrals $\{\int_{\mathfrak{G}} f_n d\lambda\}$ is a Cauchy sequence, and hence converges in $\mathbb{C}_p$. We can therefore define the integral as

$$\int_{\mathfrak{G}} f d\lambda = \lim_{n \to \infty} \int_{\mathfrak{G}} f_n d\lambda.$$

Writing $M_\lambda(f) = \int_\mathfrak{G} f d\lambda$, we get a linear functional on $C(\mathfrak{G}, \mathbb{C}_p)$ satisfying

$$|M_\lambda(f)|_p \leq \|f\|. \tag{3.4}$$

It is clear from (3.3) that if $M_{\lambda_1} = M_{\lambda_2}$, then $\lambda_1 = \lambda_2$. Finally, $M_\lambda(f)$ belongs to $\mathbb{Q}_p$ when $f$ takes values in $\mathbb{Q}_p$. Conversely, we note that every linear functional $L$ on $C(\mathfrak{G}, \mathbb{C}_p)$ satisfying $|L(f)|_p \leq \|f\|$ for all continuous $f$ and $L(f)$ belongs to $\mathbb{Q}_p$ when $f$ takes values in $\mathbb{Q}_p$, must be of the form $L = M_\lambda$ for a unique $\lambda$ in $\Lambda(\mathfrak{G})$. Indeed, the element $\lambda$ can be obtained as follows. For each open subgroup $\mathfrak{H}$ of $\mathfrak{G}$, and each coset $x$ of $\mathfrak{G}/\mathfrak{H}$, we put $c_\mathfrak{H}(x) = L(\varepsilon_x)$ where $\varepsilon_x$ is the characteristic function of $x$, and then define $\lambda_\mathfrak{H}$ by the formula (3.1). These elements $\lambda_\mathfrak{H}$ are clearly compatible and so give an element in $\Lambda_\mathfrak{G}$,

The following remarks whose proofs we omit are also useful. If $\lambda = g$ in $\mathfrak{G}$, then $dg$ is the Dirac measure given by

$$\int_\mathfrak{G} f dg = f(g).$$

Secondly, the product in $\Lambda(\mathfrak{G})$ corresponds to the convolution $*$ of measures which we recall is defined by

$$\int_\mathfrak{G} f(x) d(\lambda_1 * \lambda_2)(x) = \int_\mathfrak{G} \left( \int_\mathfrak{G} f(x + y) d\lambda_1(x) \right) d\lambda_2(y).$$

Thirdly, if $\nu : \mathfrak{G} \longrightarrow \mathbb{C}_p^\times$ is a continuous group homomorphism. then one sees easily that we can extend $\nu$ to a continuous algebra homomorphism,

$$\nu : \Lambda(\mathfrak{G}) \longrightarrow \mathbb{C}_p$$

by the formula $\nu(\lambda) = \int_\mathfrak{G} \nu d\lambda$.

Finally, to take account of the fact that the $p$-adic analogue of the complex Riemann zeta function also has a pole, we now introduce the notion of a $p$-adic pseudo-measure on $\mathfrak{G}$ [Se]. Let $Q(\mathfrak{G})$ be the total ring of fractions of $\Lambda(\mathfrak{G})$, i.e. the set of all quotients $\alpha/\beta$ with $\alpha$, $\beta$ in $\Lambda(\mathfrak{G})$ and $\beta$ a non-zero divisor. We say that an element $\lambda$ of $Q(\mathfrak{G})$ is a *pseudo-measure* on $\mathfrak{G}$ if $(g-1)\lambda$ is in $\Lambda(\mathfrak{G})$ for all $g$ in $\mathfrak{G}$. Suppose that $\lambda$ is a pseudo-measure on $\mathfrak{G}$ and let $\nu$ be a homomorphism from $\mathfrak{G}$ to $\mathbb{C}_p^\times$ which is not identically one. We can then define

$$\int_\mathfrak{G} \nu d\lambda = \frac{\int_\mathfrak{G} \nu d((g-1)\lambda)}{\nu(g) - 1},$$

where $g$ is any element of $\mathfrak{G}$ with $\nu(g) \neq 1$. This is independent of the choice of $g$ because, as remarked earlier $\nu$ extends to a ring homomorphism from $\Lambda(\mathfrak{G})$ to $\mathbb{C}_p$.

## 3.3 The Mahler Transform

The key tool for relating the ring $R = \mathbb{Z}_p[[T]]$ of formal power series studied in the previous chapter, to the Iwasawa algebra of the Galois group $\mathcal{G} = \mathrm{Gal}(\mathcal{K}_\infty/\mathbb{Q}_p)$ is provided by the following remarkable theorem of Mahler [M], whose proof we omit. As usual we define $\binom{x}{n}$ to be 1 if $n = 0$, and

$$\binom{x}{n} = \frac{x(x-1)\cdots(x-n+1)}{n!} \quad (n \geq 1).$$

**Theorem 3.3.1.** *Let* $f : \mathbb{Z}_p \longrightarrow \mathbb{C}_p$ *be any continuous function. Then* $f$ *can be written uniquely in the form*

$$f(x) = \sum_{n=0}^{\infty} a_n \binom{x}{n}, \qquad (3.5)$$

*where* $a_n \in \mathbb{C}_p$ *tends to zero as* $n \longrightarrow \infty$.

Note that the coefficients $a_n$ are given by $a_n = (\nabla^n f)(0)$ where $\nabla f(x) = f(x+1) - f(x)$. Since $\left|\binom{x}{n}\right|_p \leq 1$ for all $x$ in $\mathbb{Z}_p$, it follows that $\|f\| = \sup |a_n|_p$. If $\lambda$ is any element of $\Lambda(\mathbb{Z}_p)$, it follows from (3.4) that

$$c_n(\lambda) = \int_{\mathbb{Z}_p} \binom{x}{n} d\lambda \quad (n \geq 0) \qquad (3.6)$$

lies in $\mathbb{Z}_p$. This leads to the following definition.

**Definition 3.3.2.** *We define the Mahler transform* $\mathcal{M} : \Lambda(\mathbb{Z}_p) \longrightarrow R$ *by*

$$\mathcal{M}(\lambda) = \sum_{n=0}^{\infty} c_n(\lambda) T^n,$$

*where* $c_n(\lambda)$ *is given by* (3.6) *for* $\lambda$ *in* $\Lambda(\mathbb{Z}_p)$.

**Theorem 3.3.3.** *The Mahler transform is an isomorphism of* $\mathbb{Z}_p$-*algebras.*

*Proof.* It is clear from Theorem 3.3.1 that $\mathcal{M}$ is injective, and is a $\mathbb{Z}_p$-module homomorphism. To see that it is bijective, we construct an inverse $\Upsilon : R \longrightarrow \Lambda(\mathbb{Z}_p)$ as follows. Let $g(T) = \sum_{n=0}^{\infty} c_n T^n$ be any element of $R$. We can then define a linear functional $L$ on $C(\mathbb{Z}_p, \mathbb{C}_p)$ by

$$L(f) = \sum_{n=0}^{\infty} a_n c_n,$$

where $f$ has Mahler expansion as in (3.5) above. Of course, the series on the right converges because $a_n$ tends to zero as $n \longrightarrow \infty$. Since the $c_n$ lie in $\mathbb{Z}_p$, it is clear that $|L(f)|_p \leq \|f\|$ for all $f$. Hence there exists $\lambda$ in $\Lambda(\mathbb{Z}_p)$ such that $L = M_\lambda$, and we define $\Upsilon(g(T)) = \lambda$. It is plain that $\Upsilon$ is an inverse of $\mathcal{M}$. In fact, it can also be shown that $\mathcal{M}$ preserves products, although we omit the proof here.                                      $\square$

**Lemma 3.3.4.** *We have* $\mathcal{M}(1_{\mathbb{Z}_p}) = 1 + T$, *and thus* $\mathcal{M} : \Lambda(\mathbb{Z}_p) \longrightarrow R$ *is the unique isomorphism of topological $\mathbb{Z}_p$-algebras which sends the topological generator $1_{\mathbb{Z}_p}$ of $\mathbb{Z}_p$ to $(1 + T)$.*

*Proof.* Take $\lambda = 1_{\mathbb{Z}_p}$. By definition,

$$\mathcal{M}(\lambda) = \sum_{n=0}^{\infty} c_n(\lambda)T^n,$$

where

$$c_n(\lambda) = \int_{\mathbb{Z}_p} \binom{x}{n} d\lambda = \binom{1}{n}$$

whence the first assertion is clear. For the second assertion, we note that it is well-known (see for example, [Se1]), that for each choice of a topological generator $\gamma$ of $\mathbb{Z}_p$, there is a unique topological isomorphism of $\mathbb{Z}_p$-algebras, which maps $\gamma$ to $(1 + T)$.                    $\square$

**Lemma 3.3.5.** *For all $g$ in $R$, and all integers $k \geq 0$, we have the integral*

$$\int_{\mathbb{Z}_p} x^k d(\Upsilon(g(T))) = (D^k g(T))_{T=0}$$

*where* $D = (1 + T)\frac{d}{dT}$.

*Proof.* For fixed $g(T) = \sum_{n=0}^{\infty} c_n T^n$ in $R$, consider the linear functional $L$ on $C(\mathbb{Z}_p, \mathbb{C}_p)$ defined by

$$L(f) = \int_{\mathbb{Z}_p} x f(x) d\Upsilon(g(T)).$$

Clearly, we have $|L(f)|_p \leq \|f\|$, and so $L = M_\lambda$ for some $\lambda$ in $\Lambda(\mathbb{Z}_p)$, whence we obtain

$$\int_{\mathbb{Z}_p} x f(x) d\Upsilon(g(T)) = \int_{\mathbb{Z}_p} f(x) d\lambda. \tag{3.7}$$

We first claim that

$$\mathcal{M}(\lambda) = Dg(T). \qquad (3.8)$$

To prove this, we note that

$$Dg(T) = \sum_{n=0}^{\infty}(nc_n + (n+1)c_{n+1})T^n.$$

On the other hand, by definition, $\mathcal{M}(\lambda) = \sum_{n=0}^{\infty} e_n T^n$, where

$$e_n = \int_{\mathbb{Z}_p} x\binom{x}{n} d\Upsilon(g(T)).$$

But we have the identity

$$x\binom{x}{n} = (n+1)\binom{x}{n+1} + n\binom{x}{n} \qquad (n \geq 0),$$

whence we get $e_n = nc_n + (n+1)c_{n+1}$ for all $n \geq 0$, thereby proving (3.8). But, for all $h(T)$ in $R$, we have

$$\int_{\mathbb{Z}_p} d\Upsilon(h(T)) = h(0).$$

So the assertion of the lemma is equivalent to

$$\int_{\mathbb{Z}_p} x^k d\Upsilon(g(T)) = \int_{\mathbb{Z}_p} d\Upsilon(D^k g(\,T)) \qquad (k \geq 0).$$

By an induction argument, we have

$$\int_{\mathbb{Z}_p} d\Upsilon(D^k g(T)) = \int_{\mathbb{Z}_p} x^{k-1} d(\Upsilon(Dg(T))).$$

It is now plain by (3.8) and (3.7) that this is equal to

$$\int_{\mathbb{Z}_p} x^k d\Upsilon(g(T))$$

and the proof of the lemma is complete. □

## 3.4 Restriction of Measures

As the multiplicative group $\mathbb{Z}_p^{\times}$ is not a subgroup of the additive group $\mathbb{Z}_p$, it is intuitively surprising that we can nevertheless canonically identify $\Lambda(\mathbb{Z}_p^{\times})$ with a subset of $\Lambda(\mathbb{Z}_p)$. The aim of this section is to explain this identification and its interpretation in terms of power series.

Let $\varepsilon$ be the characteristic function of $\mathbb{Z}_p^\times$ in $\mathbb{Z}_p$. It is continuous because $\mathbb{Z}_p^\times$ is open and closed in $\mathbb{Z}_p$. Given $\lambda$ in $\Lambda(\mathbb{Z}_p)$, we can define a functional $L$ on $C(\mathbb{Z}_p, \mathbb{C}_p)$ by

$$L(f) = \int_{\mathbb{Z}_p} f\varepsilon \, d\lambda,$$

and clearly $|L(f)|_p \leq \|f\|$. Hence $L = M_{\#(\lambda)}$ for a unique $\#(\lambda)$ in $\Lambda(\mathbb{Z}_p)$. In order to interpret this operation in terms of power series, we define the operator $\mathcal{S} : R \longrightarrow R$ by

$$\mathcal{S}(g(T)) = g(T) - \frac{1}{p} \sum_{\xi \in \mu_p} g(\xi(1+T) - 1)).$$

We recall that the operator $\psi : R \longrightarrow R$ is defined in Proposition 2.2.3 of Chapter 2.

**Lemma 3.4.1.** *For all $\lambda$ in $\Lambda(\mathbb{Z}_p)$, we have $\mathcal{S}(\mathcal{M}(\lambda)) = \mathcal{M}(\#(\lambda))$. In particular, $\#(\lambda) = \lambda$ if and only if $\mathcal{S}(\mathcal{M}(\lambda)) = \mathcal{M}(\lambda)$, or equivalently if and only if $\mathcal{M}(\lambda)$ belongs to $R^{\psi=0}$.*

*Proof.* For each $n \geq 0$, let

$$\mathrm{pr}_n : \Lambda(\mathbb{Z}_p) \longrightarrow \mathbb{Z}_p[\mathbb{Z}_p/p^n\mathbb{Z}_p]$$

be the natural map. Suppose

$$\mathrm{pr}_n(\lambda) = \sum_{k=0}^{p^n-1} e_n(k)(k + p^n\mathbb{Z}_p) \tag{3.9}$$

with the $e_n(k)$ in $\mathbb{Z}_p$. Thus, by (3.3),

$$e_n(k) = \int_{\mathbb{Z}_p} \varepsilon_{k+p^n\mathbb{Z}_p} d\lambda,$$

where $\varepsilon_{k+p^n\mathbb{Z}_p}$ denotes the characteristic function of the subset $k+p^n\mathbb{Z}_p$ of $\mathbb{Z}_p$. It is then clear that $\#(\lambda)$ is the unique element of $\Lambda(\mathbb{Z}_p)$ defined by

$$\mathrm{pr}_n(\#(\lambda)) = \sum_{\substack{k=0 \\ (k,p)=1}}^{p^n-1} e_n(k)(k + p^n\mathbb{Z}_p) \quad (n \geq 1).$$

Recall that the Weierstrass preparation theorem shows that

$$\mathbb{Z}_p[T]/\omega_n\mathbb{Z}_p[T] \simeq R/\omega_n R,$$

where $\omega_n(T) = (1 + T)^{p^n} - 1$. Hence, as $R = \varprojlim R/\omega_n R$, we obtain natural maps

$$\mathrm{pr}'_n : R \longrightarrow \mathbb{Z}_p[T]/\omega_n \mathbb{Z}_p[T].$$

Since $\mathcal{M}(1_{\mathbb{Z}_p}) = 1 + T$, it follows that

$$\mathrm{pr}'_n(\mathcal{M}(\lambda)) = \sum_{k=0}^{p^n-1} e_n(k)(1 + T)^k \mod \omega_n \mathbb{Z}_p[T]$$

and that

$$\mathrm{pr}'_n(\mathcal{M}(\#\lambda)) = \sum_{\substack{k=0 \\ (k,p)=1}}^{p^n-1} e_n(k)(1 + T)^k \mod \omega_n \mathbb{Z}_p[T]$$

for all $n \geq 0$. But

$$\mathcal{S}\left( \sum_{k=0}^{p^n-1} e_n(k)(1 + T)^k \right) = \sum_{\substack{k=0 \\ (k,p)=1}}^{p^n-1} e_n(k)(1 + T)^k \quad (n \geq 1),$$

whence it is plain that $\mathcal{S}(\mathcal{M}(\lambda) = \mathcal{M}(\#(\lambda))$. The final assertion is clear from (2.2) and this completes the proof.  □

We can define a natural inclusion

$$i : \Lambda(\mathbb{Z}_p^\times) \longrightarrow \Lambda(\mathbb{Z}_p)$$

by the formula

$$\int_{\mathbb{Z}_p} f \, d(i(\eta)) = \int_{\mathbb{Z}_p^\times} f|_{\mathbb{Z}_p^\times} \, d\eta,$$

where $f$ runs over all continuous $\mathbb{C}_p$-valued functions on $\mathbb{Z}_p$, and $f|_{\mathbb{Z}_p^\times}$ denotes the restriction to $\mathbb{Z}_p^\times$.

**Lemma 3.4.2.** *We have* $i(\Lambda(\mathbb{Z}_p^\times)) = \{\lambda \in \Lambda(\mathbb{Z}_p) : \#(\lambda) = \lambda\}$. *In particular, we have* $\mathcal{M}(i(\Lambda(\mathbb{Z}_p^\times))) = R^{\psi=0}$.

*Proof.* It is clear that the image of $i$ is contained in the set on the right. Conversely, if we have a $\lambda$ in $\Lambda(\mathbb{Z}_p)$ with $\#(\lambda) = \lambda$, we obtain an element $\eta$ in $\Lambda(\mathbb{Z}_p^\times)$ by specifying that

$$\int_{\mathbb{Z}_p^\times} h \, d\eta = \int_{\mathbb{Z}_p} \tilde{h} \, d\lambda,$$

where $h$ is any continuous $\mathbb{C}_p$-valued function on $\mathbb{Z}_p^\times$, and $\tilde{h}$ denotes its extension by zero to $\mathbb{Z}_p$. Clearly $i(\eta) = \lambda$ because $\#(\lambda) = \lambda$, and the proof is complete.  □

In what follows, we suppress the map $i$ and identify $\Lambda(\mathbb{Z}_p^\times)$ with the subset of $\Lambda(\mathbb{Z}_p)$ consisting of all elements $\lambda$ with $\#(\lambda) = \lambda$.

## 3.5 The Fundamental Exact Sequence

Our goal now is to combine the above interpretation of $\Lambda(\mathbb{Z}_p^\times)$ with Theorem 2.5.2 of Chapter 2, to obtain the fundamental exact sequence needed for the proof of Iwasawa's theorem.

We consider the action of $\mathcal{G}$ on $\Lambda(\mathbb{Z}_p)$ defined by

$$g \cdot (a_{\mathbb{Z}_p}) = (\chi(g) \cdot a)_{\mathbb{Z}_p} \qquad (a \in \mathbb{Z}_p),$$

where we write $a_{\mathbb{Z}_p}$ to stress that we are viewing $a$ as an element of the group $\mathbb{Z}_p$ in the Iwasawa algebra $\Lambda(\mathbb{Z}_p)$. By linearity and continuity, this extends to an action of $\mathcal{G}$ on $\Lambda(\mathbb{Z}_p)$, and $\Lambda(\mathbb{Z}_p^\times)$ is then a $\mathcal{G}$-submodule. Moreover, since $\mathcal{M}(1_{\mathbb{Z}_p}) = 1 + T$, it is clear that $\mathcal{M}$ is a $\mathcal{G}$-isomorphism from $\Lambda(\mathbb{Z}_p)$ to $R$ when $R$ is endowed with the $\mathcal{G}$-action given by (2.7). Finally, we note that there is a canonical $\mathcal{G}$-isomorphism

$$\tilde{\chi} : \Lambda(\mathcal{G}) \simeq \Lambda(\mathbb{Z}_p^\times) \tag{3.10}$$

induced by the isomorphism $\chi : \mathcal{G} \simeq \mathbb{Z}_p^\times$ given by the cyclotomic character. Let

$$\widetilde{\mathcal{M}} : \Lambda(\mathcal{G}) \simeq R^{\psi=0}$$

be the $\mathcal{G}$-isomorphism defined by $\widetilde{\mathcal{M}} = \mathcal{M} \circ \tilde{\chi}$. Recall that $\mathcal{U}_\infty$ denotes the projective limit of the local units with respect to the norm maps in the cyclotomic tower, endowed with its natural action of $\mathcal{G}$. Define

$$\tilde{\mathcal{L}} : \mathcal{U}_\infty \longrightarrow \Lambda(\mathcal{G})$$

by

$$\tilde{\mathcal{L}}(\mathbf{u}) = \widetilde{\mathcal{M}}^{-1}(\mathcal{L}(\mathbf{f_u})),$$

where $\mathbf{f_u}$ denotes the Coleman power series of $\mathbf{u}$ and $\mathcal{L}$ is given by (2.10). It is a $\mathcal{G}$-homomorphism, and clearly

$$\mathcal{L}(\mathbf{f_u}) = \sum_{n=0}^{\infty} T^n \int_{\mathcal{G}} \binom{\chi(g)}{n} d\tilde{\mathcal{L}}(\mathbf{u}).$$

The following theorem was first proven in [Co2] (see also [Sa], and [O] for closely related sequences).

**Theorem 3.5.1.** *We have an exact sequence of $\mathcal{G}$-modules*

$$0 \longrightarrow \mu_{p-1} \times T_p(\mu) \longrightarrow \mathcal{U}_\infty \xrightarrow{\tilde{\mathcal{L}}} \Lambda(\mathcal{G}) \xrightarrow{\beta} T_p(\mu) \longrightarrow 0, \tag{3.11}$$

*where the kernel on the left is the natural inclusion, and the map $\beta$ on the right is given by $\beta(\lambda) = (\zeta_n)^{\int_\mathcal{G} \chi d\lambda}$.*

*Proof.* This is none other than a reinterpretation of the exact sequence
(2.11) of Theorem 2.5.2. Specifically, by construction, we have the commutative diagram

$$
\begin{array}{ccc}
\mathcal{U}_\infty & \xrightarrow{\;\tilde{\mathcal{L}}\;} & \Lambda(\mathcal{G}) \\
\downarrow & & \downarrow{\widetilde{\mathcal{M}}} \\
W & \xrightarrow{\;\mathcal{L}\;} & R^{\psi=0}
\end{array}
$$

where the left vertical arrow is the $\mathcal{G}$-isomorphism $\mathbf{u} \mapsto f_{\mathbf{u}}(T)$, and as
noted above, $\widetilde{\mathcal{M}}$ is also a $\mathcal{G}$-isomorphism. If $x$ is any element of $\mu_{p-1}$, and
$a$ is any element of $\mathbb{Z}_p$, then $\mathbf{u} = (x\zeta_n^a)$ is in $\mathcal{U}_\infty$, and $\mathbf{u}(T) = x(1+T)^a$.
Finally the maps $\alpha$ and $\beta$ are compared by using Lemma 3.3.5 and the
isomorphism (3.10). This completes the proof.    □

We recall the logarithmic derivative homomorphims $\delta_k$ defined by
(2.12).

**Proposition 3.5.2.** *For all $k \geq 1$, and all $\mathbf{u}$ in $\mathcal{U}_\infty$, we have*

$$
\int_{\mathcal{G}} \chi(g)^k d\tilde{\mathcal{L}}(\mathbf{u}) = (1 - p^{k-1})\delta_k(\mathbf{u}). \tag{3.12}
$$

*Proof.* We first note that, for any $\lambda$ in $\Lambda(\mathcal{G})$, the isomorphism $\tilde{\chi}$ given
by (3.10) gives

$$
\int_{\mathcal{G}} \chi(g)^k d\lambda = \int_{\mathbb{Z}_p^\times} x^k d(\tilde{\chi}(\lambda)).
$$

Also, via our identification of $\Lambda(\mathbb{Z}_p^\times)$ with a subset of $\Lambda(\mathbb{Z}_p)$, the integral
above on the right has the same value if we integrate over the whole
of $\mathbb{Z}_p$. Now take $\lambda = \tilde{\mathcal{L}}(\mathbf{u})$, so that, by definition, we have $\tilde{\chi}(\tilde{\mathcal{L}}(\mathbf{u})) = \Upsilon(\mathcal{L}(f_{\mathbf{u}}))$, where we recall that

$$
\mathcal{L}(f_{\mathbf{u}})(T) = \frac{1}{p} \log\left(\frac{f_{\mathbf{u}}(T)^p}{\varphi(f)(T)}\right).
$$

Thus, by Lemma 3.3.5, the integral on the left of (3.12) is equal to

$$
\left(D^{k-1}(h_{\mathbf{u}}(T) - \varphi(h_{\mathbf{u}})(T))\right)_{T=0},
$$

where

$$
h_{\mathbf{u}}(T) = (1+T)\frac{f_{\mathbf{u}}'(T)}{f_{\mathbf{u}}(T)}.
$$

Using (2.13), it follows that this expression is equal to the right
hand side of (3.12), and the proof of the proposition is complete.    □

## 3.6 Image of $\delta_k$

As an interesting digression, we now determine the image of all the logarithmic derivative homomorphisms $\delta_k$ using the fundamental exact sequence (3.11), and (3.12). We recall that Theorem 2.6.4 of the previous chapter determines the image of $\delta_k$ for $k = 1, \cdots, p - 1$.

**Proposition 3.6.1.** *Let $k \geq 1$ be any integer. If $k = 1$ or $k \not\equiv 1 \bmod (p-1)$, then $\delta_k(\mathcal{U}_\infty) = \mathbb{Z}_p$. If $k$ is greater than 1 and $k \equiv 1 \bmod (p-1)$, then $\delta_k(\mathcal{U}_\infty) = p^m \mathbb{Z}_p$, where $m = 1 + \mathrm{ord}_p(k-1)$.*

To prove this proposition, we need an algebraic description of $\Lambda(\mathcal{G})$ as $p - 1$ copies of the ring $R$. We stress that this second description is of a totally different nature to that given in sections 3.4 and 3.5, which is based on Mahler's theorem.

We have

$$\mathcal{G} = \varpi \times \Gamma \tag{3.13}$$

where $\varpi$ is the cyclic group of order $p - 1$, and $\Gamma$ is isomorphic to $\mathbb{Z}_p$. Write $\chi_\varpi$ for the restriction of $\chi$ to $\varpi$, so that the characters of $\varpi$ are given by the $\chi_\varpi^i$ for $i$ running over a complete set of residues modulo $p - 1$. Fix a topological generator $\gamma$ of $\Gamma$. Let $\mathcal{A}$ denote the group ring $\mathbb{Z}_p[\varpi]$. It is easily seen that $\Lambda(\mathcal{G})$ can be identified with the Iwasawa algebra of $\Gamma$ over $\mathcal{A}$ and hence that there is a unique topological isomorphism

$$\Lambda(\mathcal{G}) \simeq \mathcal{A}[[T]]$$

which is the identity on $\mathcal{A}$ and maps $\gamma$ to $1 + T$. On the other hand, we have a canonical isomorphism

$$\mathcal{A}[[T]] \simeq \bigoplus_{i \bmod (p-1)} R$$

defined by mapping a power series $f = \sum_{n=0}^{\infty} a_n T^n$ in $\mathcal{A}[[T]]$ to the tuple

$$\left( \sum_{n=0}^{\infty} \chi_\varpi^i(a_n) T^n \right)_i.$$

The composition of these two isomorphisms yields an isomorphism

$$\vartheta : \Lambda(\mathcal{G}) \xrightarrow{\sim} \bigoplus_{i \bmod (p-1)} R$$

and we write $\vartheta(\lambda)_i$ for the components of $\vartheta(\lambda)$.

**Lemma 3.6.2.** *For all $\lambda$ in $\Lambda(\mathcal{G})$ and all integers $k \geq 1$, we have*

$$\int_{\mathcal{G}} \chi(g)^k d\lambda = \vartheta(\lambda)_i(\chi(\gamma)^k - 1),$$

*where $i = k$ mod $(p-1)$.*

**Corollary 3.6.3.** $\vartheta(\mathrm{Im}(\tilde{\mathcal{L}}))$ *is the subset of* $\displaystyle\bigoplus_{i \bmod (p-1)} R$ *which consists of all tuples $(\alpha_i)$ where $\alpha_i$ runs over $R$ when $i \not\equiv 1$ mod $(p-1)$, and $\alpha_1$ runs over $((1+T) - \chi(\gamma))R$.*

*Proof.* The corollary follows immediately since the lemma shows that

$$\int \chi(g) d\lambda = 0$$

if and only if $\vartheta(\lambda)_1(T)$ vanishes at $T = \chi(\gamma) - 1$.    □

We remark that it is in no way clear arithmetically how one constructs a unit **u** in $\mathcal{U}_\infty$ such that, for example

$$\vartheta(\tilde{\mathcal{L}}(\mathbf{u}))_i = 1 \text{ for } i \not\equiv 1 \mod (p-1), \text{ and } \vartheta(\tilde{\mathcal{L}}(\mathbf{u}))_1 = 1 + T - \chi(\gamma).$$

However the above corollary shows that such **u** must exist.

*Proof of Lemma 3.6.2.* We recall that

$$\chi^k(\lambda) = \int_{\mathcal{G}} \chi^k(g) d\lambda.$$

On the other hand, writing $\lambda = \sum_{n=0}^{\infty} a_n T^n$, with $a_n \in \mathbb{Z}_p[\varpi]$, it is clear by linearity and continuity that

$$\chi^k(\lambda) = \sum_{n=0}^{\infty} \chi^k(a_n)(\chi(\gamma)^k - 1)^n = \vartheta(\lambda)_i(\chi(\gamma)^k - 1).$$

□

*Proof of Proposition 3.6.1.* We combine Lemma 3.6.2 and Proposition 3.5.2. Also, we must assume that $k > 1$, since (3.12) tells us nothing about the image of $\delta_1$ (but the image of $\delta_1$ is determined by Theorem 2.6.4 of the previous chapter). Noting $(1 - p^{k-1})$ is a $p$-adic unit since $k > 1$, it follows that the image of $\delta_k$ is equal to the subset of $\mathbb{Z}_p$ given by

$$J = \begin{cases} \{h(\chi(\gamma) - 1) : h \in R\} & \text{if } k \not\equiv 1 \mod (p-1) \\ \{(\chi(\gamma)^k - \chi(\gamma)) \cdot h(\chi(\gamma) - 1) : h \in R\} & \text{if } k \equiv 1 \mod (p-1). \end{cases}$$

When $k \not\equiv 1 \mod (p-1)$, it is clear that $J = \mathbb{Z}_p$. On the other hand, $\chi(\gamma) = 1 + pw$, with $w$ in $\mathbb{Z}_p^\times$, so it is plain that $J = p^m \mathbb{Z}_p$ when $k \equiv 1 \mod (p-1)$, where $m = \mathrm{ord}_p(k-1) + 1$. This completes the proof. $\square$

We end this chapter by remarking that a weak form of Proposition 3.6.1 can be used to give an alternative proof of Theorem 2.1.2 and the exact sequence 3.11 (see [Sa] and the original paper [CW]).

# Cyclotomic Units and Iwasawa's Theorem

## 4.1 Introduction

In this chapter, we use the fundamental exact sequence (3.11) of the previous chapter to prove both the existence of the $p$-adic analogue of the Riemann-zeta function, and Iwasawa's theorem. We remark that, even though Iwasawa did not explicitly use the language of $p$-adic measures on Galois groups, he was the first person to prove that the $p$-adic analogue of $\zeta(s)$ could be expressed in terms of what amounts to $p$-adic integrals on the Galois group of the field generated over $\mathbb{Q}$ by all $p$-power roots of unity. However, he used $p$-adic measures coming from the classical Stickelberger theorem rather than those here arising from cyclotomic units [Iw3]. We also remark that Iwasawa's original ingenious and difficult proof in [Iw2] of Theorem 4.4.1 is very different from the one given here. Our approach via the exact sequence (3.11) has the advantage of establishing all of these results simultaneously. We then discuss the relationship of Iwasawa's theorem to the main conjecture and end the chapter by proving some rather delicate facts about unit groups and ideal class groups of finite extensions of $\mathbb{Q}$ contained in $F_\infty$. Our method of proof is the classical one in Iwasawa theory of deriving assertions at finite levels from the corresponding ones for $F_\infty$. These final results are needed to complete the proof of the main conjecture via Euler systems in Chapter 6.

## 4.2 $p$-adic Zeta Function

We first establish the existence of the $p$-adic analogue of the Riemann zeta function. Recall that $\mathcal{F}_\infty = \mathbb{Q}(\mu_{p^\infty})$ and that $F_\infty = \mathcal{F}_\infty^+$ is its maximal real subfield. Let

$$\mathcal{G} = \mathrm{Gal}(\mathcal{F}_\infty/\mathbb{Q}), \quad G = \mathrm{Gal}(F_\infty/\mathbb{Q}).$$

We shall often need the following elementary lemma, which enables us to identify $\Lambda(G)$ with a sub-algebra of $\Lambda(\mathcal{G})$. Let

$$\mathcal{J} = \{1, \iota\} = \mathrm{Gal}(\mathcal{F}_\infty/F_\infty).$$

If $M$ is any $\mathbb{Z}_p[\mathcal{J}]$-module, then since $p$ is odd, there is the decomposition

$$M = M^+ \oplus M^-, \text{ where } M^+ = \frac{1+\iota}{2}M, \ M^- = \frac{1-\iota}{2}M.$$

In particular, we have

$$\Lambda(\mathcal{G}) = \Lambda(\mathcal{G})^+ \oplus \Lambda(\mathcal{G})^-. \tag{4.1}$$

**Lemma 4.2.1.** *The restriction to $\Lambda(\mathcal{G})^+$ of the natural surjection from $\Lambda(\mathcal{G})$ onto $\Lambda(G)$ induces an isomorphism*

$$\Lambda(\mathcal{G})^+ \simeq \Lambda(G). \tag{4.2}$$

*Proof.* Recall that

$$\mathcal{F}_n = \mathbb{Q}(\mu_{p^{n+1}}), \quad F_n = \mathbb{Q}(\mu_{p^{n+1}})^+, \tag{4.3}$$

and write $\mathcal{G}_n = \mathrm{Gal}(\mathcal{F}_n/\mathbb{Q}), \ G_n = \mathrm{Gal}(F_n/\mathbb{Q})$. Let

$$\pi_n : \mathbb{Z}_p[\mathcal{G}_n] \longrightarrow \mathbb{Z}_p[G_n]$$

denote the natural surjection. We claim that $\pi_n$ induces an isomorphism from $\mathbb{Z}_p[\mathcal{G}_n]^+$ onto $\mathbb{Z}_p[G_n]$. Indeed, it is clear that $\pi_n$ is surjective, and that it maps $\mathbb{Z}_p[\mathcal{G}_n]^-$ to zero. To complete the proof, we note that the $\mathbb{Z}_p$-rank of $\mathbb{Z}_p[\mathcal{G}_n]^+$ is equal to $((p-1)/2)p^n$, because

$$\mathbb{Z}_p[\mathcal{G}_n]^+ = \bigoplus_{\substack{i \text{ even} \\ i \bmod p-1}} \mathbb{Z}_p[\mathcal{G}_n]^{(i)}$$

where the sum on the right is taken over the eigenspaces for the even powers modulo $(p-1)$ of the character giving the action of $\mathcal{G}_0$ on $\mu_p$. The assertion of the lemma now follows on passing to the projective limit. $\square$

From now on, we shall identify $\Lambda(G)$ with the subalgebra $\Lambda(\mathcal{G})^+$ of $\Lambda(\mathcal{G})$.

**Lemma 4.2.2.** *Assume that $\lambda$ is any element of $\Lambda(\mathcal{G})$ such that $\chi^k(\lambda) = 0$ for all $k > 0$. Then $\lambda = 0$. The analogous assertion is also valid for pseudo-measures on $\mathcal{G}$.*

*Proof.* Recall from section 3.5 the isomorphism

$$\widetilde{\mathcal{M}} : \Lambda(\mathcal{G}) \simeq R^{\psi=0}$$

arising from Mahler's theorem. Hence

$$\widetilde{\mathcal{M}}(\lambda) = g(T),$$

where

$$g(T) = \sum_{n=0}^{\infty} T^n \int_{\mathcal{G}} \binom{\chi(g)}{n} d\lambda.$$

But when $n > 0$, the binomial coefficient $\binom{x}{n}$ is a polynomial in $x$ with constant term equal to zero. Hence by the hypothesis of the lemma, we have

$$\int_{\mathcal{G}} \binom{\chi(g)}{n} d\lambda = 0 \quad (n > 0).$$

Thus $g(T)$ is a constant and therefore must be identically zero since it belongs to $R^{\psi=0}$. This completes the proof of the lemma for elements of $\Lambda(\mathcal{G})$.

A consequence is that if $\lambda$ is any element of $\Lambda(\mathcal{G})$ such that $\chi^k(\lambda) \neq 0$ for all integers $k > 0$, then $\lambda$ is not a zero divisor. Indeed, if $\lambda' \cdot \lambda = 0$, then since $\chi^k$ is a ring homomorphism of $\Lambda(\mathcal{G})$, it follows that $\chi^k(\lambda') = 0$ for all $k > 0$, whence $\lambda' = 0$.

Suppose now that $\xi$ is a pseudo-measure on $\mathcal{G}$ with $\chi^k(\xi) = 0$ for all $k > 0$. For each $u \in \mathbb{Z}_p^\times$, let $\sigma_u$ denote the unique element of $\mathcal{G}$ with $\chi(\sigma_u) = u$. Now choose $u = 1 + p$. Then

$$\chi^k(\sigma_u - 1) = (1 + p)^k - 1 \neq 0$$

for all $k > 0$, so that $\sigma_u - 1$ is not a zero divisor in $\Lambda(\mathcal{G})$ by the remark above. But the hypothesis $\chi^k(\xi) = 0$ implies that

$$\chi^k((\sigma_u - 1)\xi) = 0$$

for all $k > 0$. As $\sigma_u - 1$ is not a zero divisor, the remark at the end of the previous paragraph shows that $\xi = 0$. $\qquad \square$

Let $Q(\mathcal{G})$ be the total ring of quotients of $\Lambda(\mathcal{G})$. Parallel to (4.1), we have a decomposition

$$Q(\mathcal{G}) = Q(\mathcal{G})^+ \oplus Q(\mathcal{G})^-$$

and it is then easy to see that one can identify pseudo-measures on $G$ with pesudo-measures on $\mathcal{G}$ which lie in $Q(\mathcal{G})^+$.

**Corollary 4.2.3.** *Let* $\lambda$ *be an element of* $\Lambda(\mathcal{G})$. *If* $\int_\mathcal{G} \chi^k d\lambda = 0$ *for* $k = 1, 3, 5, \cdots$, *then* $\lambda \in \Lambda(\mathcal{G})^+$, *and if* $\int_\mathcal{G} \chi^k d\lambda = 0$ *for* $k = 2, 4, 6, \cdots$, *then* $\lambda \in \Lambda(\mathcal{G})^-$. *The analogous assertion holds for pseudo-measures on* $\mathcal{G}$.

*Proof.* Assume $\lambda$ is in $\Lambda(\mathcal{G})$, and let $\lambda = \lambda^+ + \lambda^-$ be its decomposition as in (4.1). Since $\chi(\iota) = -1$, it is then clear that $\chi^k(1 + \iota) = 0$ for all odd integers $k$, and $\chi^k(1 - \iota) = 0$ for all even integers $k$. The assertion is then clear from the preceding lemma. A similar argument holds for pseudo-measures. □

The following proposition is the crucial one in proving the existence of the $p$-adic zeta function.

**Proposition 4.2.4.** *There exists a unique pseudo-measure* $\tilde{\zeta}_p$ *on* $\mathcal{G}$ *such that*

$$\int_\mathcal{G} \chi(g)^k d\tilde{\zeta}_p = \begin{cases} (1 - p^{k-1})\zeta(1 - k) & \text{if } k = 2, 4, \cdots \\ 0 & \text{if } k = 1, 3, \cdots. \end{cases} \tag{4.4}$$

*Proof.* The uniqueness is clear from the preceding lemmas, and the subtle part of the proof is the existence. Let $a$ and $b$ be integers which are prime to $p$ and such that $b \neq \pm a$. As before, define $\mathbf{c}(a, b) = \{c_n(a, b)\}$ where

$$c_n(a, b) = \frac{\zeta_n^{-a/2} - \zeta_n^{a/2}}{\zeta_n^{-b/2} - \zeta_n^{b/2}}.$$

As remarked earlier, $\mathbf{c}(a, b)$ is in $\mathcal{U}_\infty$. Define $\lambda(a, b)$ in $\Lambda(\mathcal{G})$ by

$$\lambda(a, b) = \tilde{\mathcal{L}}(\mathbf{c}(a, b)). \tag{4.5}$$

By Propositions 2.6.3 and 3.5.2, we have

$$\int_\mathcal{G} \chi^k d\lambda(a, b) = \begin{cases} (b^k - a^k)(1 - p^{k-1})\zeta(1 - k) & \text{if } k = 2, 4, \cdots \\ 0 & \text{if } k = 1, 3, \cdots. \end{cases} \tag{4.6}$$

Define the following element in $\Lambda(\mathcal{G})$

$$\theta(a,b) = \sigma_b - \sigma_a, \tag{4.7}$$

where, as before, $\sigma_u$ denotes the unique element of $\mathcal{G}$ with $\chi(\sigma_u) = u$ for each $u$ in $\mathbb{Z}_p^\times$. Thus, for each integer $k > 0$, we have

$$\chi^k(\theta(a,b)) = b^k - a^k \neq 0$$

because $b \neq \pm a$. Hence $\theta(a,b)$ is not a zero divisor in $\Lambda(\mathcal{G})$, and so

$$\tilde{\zeta}_p = \frac{\lambda(a,b)}{\theta(a,b)}$$

lies in $Q(\mathcal{G})$. We claim that $\tilde{\zeta}_p$ is independent of the pair $(a,b)$. Indeed, if $(a',b')$ is a second choice, then it is clear from (4.6) that

$$\chi^k(\theta(a',b')\lambda(a,b)) = \chi^k(\theta(a,b)\lambda(a',b'))$$

for all integers $k > 0$, whence it follows from Lemma 4.2.2 that

$$\theta(a',b')\lambda(a,b) = \theta(a,b)\lambda(a',b').$$

This establishes the independence. To show that $\tilde{\zeta}_p$ is a pseudo-measure, we use Lemma 4.2.5 below. We take $a = e$, $b = 1$, where $e$ is a primitive root modulo $p$ with $e^{p-1} \not\equiv 1 \bmod p^2$. By Lemma 4.2.5, the augmentation ideal $I(\mathcal{G})$ of $\Lambda(\mathcal{G})$ is generated by $\theta(e,1)$. But if $\sigma$ is any element of $\mathcal{G}$, then $\sigma - 1$ belongs to $I(\mathcal{G})$ and so $\sigma - 1 = \theta(e,1)\lambda$ for some $\lambda$ in $\Lambda(\mathcal{G})$. Therefore it is clear that $(\sigma - 1)\tilde{\zeta}_p$ belongs to $\Lambda(\mathcal{G})$ as required. This completes the proof of the proposition. $\qquad\square$

**Lemma 4.2.5.** *Let $e$ be a primitive root modulo $p$ such that $e^{p-1} \not\equiv 1$ $\bmod p^2$. Let $I(\mathcal{G})$ be the augmentation ideal of $\Lambda(\mathcal{G})$. Then*

$$I(\mathcal{G}) = (\sigma_e - 1)\Lambda(\mathcal{G}) = \Lambda(\mathcal{G})\theta(e,1).$$

*Proof.* All is clear on noting that $\sigma_e$ is a topological generator of $\mathcal{G}$, and if $K$ is any finite cyclic group, then the augmentation ideal of $\mathbb{Z}_p[K]$ is $(\tau - 1)\mathbb{Z}_p[K]$ where $\tau$ is any generator of $K$. $\qquad\square$

The following theorem is an immediate consequence of (4.4) and Corollary 4.2.3.

**Theorem 4.2.6.** *There exists a unique pseudo-measure $\zeta_p$ on $G$ such that*

$$\int_G \chi(g)^k d\zeta_p = (1 - p^{k-1})\zeta(1 - k) \tag{4.8}$$

*for all even integers $k \geq 2$.* $\qquad\square$

The pseudo-measure $\zeta_p$ occurring in this theorem is of course our desired $p$-adic analogue of the complex Riemann zeta function.

## 4.3 Cyclotomic Units

In this section, we recall the classical definition of the cyclotomic units for the fields $\mathcal{F}_n$ and $F_n$, and several basic results about them. For detailed proofs of these assertions, see [Iw2] and [Si]. As always, we let $\pi_n = \zeta_n - 1$.

**Definition 4.3.1.** *For each $n \geq 0$, we define $\mathcal{D}_n$ to be the intersection of the group of global units of $\mathcal{F}_n$ with the subgroup of $\mathcal{F}_n^\times$ generated by the $\sigma(\pi_n)$ where $\sigma$ runs over the elements of $\mathrm{Gal}(\mathcal{F}_n/\mathbb{Q})$. We also define $D_n = \mathcal{D}_n \cap F_n$.*

In fact, it is well known and easily seen that $D_n$ is generated by all the Galois conjugates of $\pm c_n(e, 1)$, where we recall that

$$c_n(e, 1) = \frac{\zeta_n^{-e/2} - \zeta_n^{e/2}}{\zeta_n^{-1/2} - \zeta_n^{1/2}} \, ,$$

and the integer $e$ is a primitive root modulo $p$ and $e^{p-1} \not\equiv 1 \bmod p^2$. Although it will not be needed for the proof of Iwasawa's theorem given in the next section, we recall the classical result that $D_n$ has finite index in the group of all units of $F_n$, and that this index is equal to the class number of $F_n$.

Recall that we have the local fields

$$\mathcal{K}_n = \mathbb{Q}_p(\mu_{p^{n+1}}), \quad K_n = \mathbb{Q}_p(\mu_{p^{n+1}})^+.$$

If $A$ is any subset of these fields, then $\overline{A}$ will denote its closure in the $p$-adic topology.

**Definition 4.3.2.** *We define $\mathcal{C}_n = \overline{\mathcal{D}_n}$ and $C_n = \overline{D_n}$.*

Recall that $\mathcal{U}_n$ denotes the group of units of $\mathcal{K}_n$ and denote by $\mathfrak{p}_n$ the maximal ideal of the ring of integers of $\mathcal{K}_n$. Similarly, denote by $U_n$ the group of units of $K_n$.

**Definition 4.3.3.** *We denote by $\mathcal{U}_n^1$ the subgroup $\{x \in \mathcal{U}_n : x \equiv 1 \bmod \mathfrak{p}_n\}$. More generally, if $Z$ is any subgroup of $\mathcal{U}_n$, we write $Z^1 = Z \cap \mathcal{U}_n^1$.*

Note that the index of $Z^1$ in $Z$ always divides $p-1$, because $Z^1$ is the kernel of the reduction map from $Z$ into $\mathbb{F}_p^\times$. In particular, this index is prime to $p$. Also, note that $\mathcal{U}_n^1$ and $U_n^1$ are now $\mathbb{Z}_p$-modules, whereas this is plainly not true for $\mathcal{U}_n$ and $U_n$ themselves.

We shall be particularly interested in the groups $U_n^1$ and $C_n^1$ and their projective limits

$$U_\infty^1 = \varprojlim U_n^1, \quad C_\infty^1 = \varprojlim C_n^1, \tag{4.9}$$

taken with respect to the norm maps. Since $U_n^1$ and $C_n^1$ are compact $\mathbb{Z}_p$-modules, so are $U_\infty^1$ and $C_\infty^1$. Moreover, they are endowed with a natural continuous action of $G = \mathrm{Gal}(F_\infty/\mathbb{Q})$. Hence they become modules over the Iwasawa algebra $\Lambda(G)$. As before, let $e$ be a fixed primitive root modulo $p$ such that $e^{p-1} \not\equiv 1 \bmod p^2$.

**Lemma 4.3.4.** *We have $C_\infty^1 = \Lambda(G)\mathbf{b}$ where $\mathbf{b} = (b_n)$ is given by*

$$b_n = uc_n(e,1) \quad (n \geq 0),$$

*and $u$ is the unique $(p-1)$-th root of unity in $\mathbb{Q}_p$ such that $eu \equiv 1 \bmod p$.*

*Proof.* Since $u^p = u$, it is clear that $\mathbf{b} = (b_n)$ belongs to $U_\infty$. Moreover, we claim that $b_n \equiv 1 \bmod \mathfrak{p}_n$ for all $n \geq 0$. Indeed, if $f_\mathbf{c}(T)$ is the Coleman power series of $\mathbf{c} = (c_n(e,1))$, then $f_\mathbf{c}(0) = e$. Hence $c_n(e,1) \equiv e \bmod \mathfrak{p}_n$ and our assertion is clear. Also, we must show that $b_n$ lies in the closure of $D_n^1$. But this is clear because $b_n^{p-1}$ certainly does lie in $D_n^1$, and $p-1$ is a $p$-adic unit. Now put $\mathfrak{b} = \Lambda(G)\mathbf{b}$ and let $h_n : C_\infty^1 \longrightarrow C_n^1$ be the natural projection. To show that $\mathfrak{b} = C_\infty^1$, it suffices to prove that

$$h_n(\mathfrak{b}) = C_n^1 \quad \text{for all } n \geq 0.$$

But $h_n(\mathfrak{b})$ is clearly the $\mathbb{Z}_p$-submodule generated by the $\sigma(b_n)$ for all $\sigma$ in $\mathrm{Gal}(F_n/\mathbb{Q})$. The assertion now follows easily from the fact that the $\pm\sigma(c_n(e,1))$ generate $D_n$ as a $\mathbb{Z}$-module. This completes the proof. $\square$

## 4.4 Iwasawa's Theorem

As stressed already in Chapter 1, the following theorem of Iwasawa is of great importance both because historically its proof led to the discovery of the Main Conjecture, and it remains today a crucial step in the proof of the Main Conjecture by our methods. Recall that $\zeta_p$ is the $p$-adic zeta function whose existence is given by Theorem 4.2.6.

**Theorem 4.4.1.** *The $\Lambda(G)$-module $U_\infty^1/C_\infty^1$ is canonically isomorphic to $\Lambda(G)/I(G)\cdot\zeta_p$ where $\zeta_p$ is the p-adic zeta function, and $I(G)$ denotes the augmentation ideal of $\Lambda(G)$.*

*Proof.* We note that, since the norm map from $K_n$ to $K_{n-1}$ induces the identity map on the residue fields, we have

$$\mathcal{U}_\infty = \mu_{p-1} \times \mathcal{U}_\infty^1, \quad U_\infty = \mu_{p-1} \times U_\infty^1.$$

Hence (3.11) can be rewritten as an exact sequence

$$0 \longrightarrow T_p(\mu) \longrightarrow \mathcal{U}_\infty^1 \overset{\mathcal{L}^1}{\longrightarrow} \Lambda(\mathcal{G}) \longrightarrow T_p(\mu) \longrightarrow 0,$$

with $\mathcal{L}^1$ being the restriction of $\widetilde{\mathcal{L}}$ to $\mathcal{U}_\infty^1$. As $p$ is odd, the above sequence remains exact after taking invariants under $J = \{1, \iota\}$. Since $T_p(\mu)^J = 0$, there is a canonical $\Lambda(G)$-isomorphism

$$\mathcal{L}^1 : U_\infty^1 \simeq \Lambda(G). \tag{4.10}$$

But by Lemma 4.3.4, we have

$$C_\infty^1 = \Lambda(G).\mathbf{b},$$

so that

$$\mathcal{L}^1(C_\infty^1) = \Lambda(G)\mathcal{L}^1(\mathbf{b}).$$

But the proof of Proposition 4.2.4 shows that $\mathcal{L}^1(\mathbf{b}) = \zeta_p\theta^+(e,1)$ where $\theta^+(e,1)$ denotes the image of $\theta(e,1)$ in $\Lambda(G)$. However, the analogue of Lemma 4.2.5 for $G$ shows that $\Lambda(G)\theta^+(e,1) = I(G)$. This completes the proof of the theorem.                                              □

## 4.5 Relation to the Main Conjecture

The $\Lambda(G)$-module $U_\infty^1/C_\infty^1$ is of interest largely because it is closely related to another $\Lambda(G)$-module of greater intrinsic arithmetic importance. We now introduce this second module, which is denoted by $X_\infty$. In fact, as is explained below, these two modules coincide in all known numerical examples, but it remains unknown whether this is true for all primes $p$. We also remark that in our proof of Iwasawa's theorem, we have made no use of local or global class field theory. However, the description of $X_\infty$ below rests crucially on global class field theory.

Recall that $F_\infty = \mathbb{Q}(\mu_{p^\infty})^+$, and define $M_\infty$ to be the maximal abelian $p$-extension of $F_\infty$, which is unramified outside of the unique prime above $p$. Put

$$X_\infty = \mathrm{Gal}(M_\infty/F_\infty). \tag{4.11}$$

As always in Iwasawa theory (see Appendix), there is a natural continuous action of $G$ on $X_\infty$ as follows. By maximality, $F_\infty$ is clearly Galois over $\mathbb{Q}$. Given any $g$ in $G$, and any $x$ in $X_\infty$, we define

$$g.x = \tilde{g}x\tilde{g}^{-1},$$

where $\tilde{g}$ denotes any lifting of $g$ to the Galois group of $M_\infty$ over $\mathbb{Q}$. As $X_\infty$ is a compact $\mathbb{Z}_p$-module, this action extends by linearity and continuity to an action of the whole Iwasawa algebra $\Lambda(G)$ on $X_\infty$. Similarly, we define $L_\infty$ to be the maximal abelian $p$-extension of $F_\infty$, which is everywhere unramified, and put

$$Y_\infty = \mathrm{Gal}(L_\infty/F_\infty). \tag{4.12}$$

Again it is clear that $L_\infty$ is Galois over $\mathbb{Q}$, and then $Y_\infty$ has a continuous action of $G$ in an entirely similar manner to the above, and becomes a $\Lambda(G)$-module. To relate these modules to $U_\infty^1/C_\infty^1$, we need the following description of $\mathrm{Gal}(M_\infty/L_\infty)$, obtained by using the full force of class field theory.

**Definition 4.5.1.** *We define $V_n$ to be the group of global units of $F_n$, and let $E_n = \overline{V_n}$.*

As before, let $E_n^1 = E_n \cap U_n^1$ and put

$$E_\infty^1 = \varprojlim E_n^1, \tag{4.13}$$

the projective limit being taken with respect to the norm maps as before.

**Theorem 4.5.2.** *We have an exact sequence of $\Lambda(G)$-modules*

$$0 \longrightarrow E_\infty^1 \longrightarrow U_\infty^1 \longrightarrow \mathrm{Gal}(M_\infty/L_\infty) \longrightarrow 0.$$

*Proof.* For each $n \geq 0$, let $M_n$ (resp. $L_n$) denote the maximal abelian $p$-extension of $F_n$ which is unramified outside $p$ (resp. unramified everywhere). A standard exercise in the fundamental theorem of global class field theory (see [Wa, Chapter 13, §13.1, Corollary 13.6]) shows that the Artin map induces a $G_n$-isomorphism

$$U_n^1/E_n^1 \simeq \mathrm{Gal}(M_n/L_n). \tag{4.14}$$

Since

$$M_\infty = \bigcup_{n \geq 0} M_n, \quad L_\infty = \bigcup_{n \geq 0} L_n,$$

we obtain the exact sequence of the theorem by passing to the projective limit over $n$ in (4.14), and noting that $\varprojlim$ is an exact functor here because all are finitely generated $\mathbb{Z}_p$-modules.                    □

In view of Theorem 4.5.2, we obtain the following four term exact sequence of $\Lambda(G)$-modules

$$0 \longrightarrow E_\infty^1/C_\infty^1 \longrightarrow U_\infty^1/C_\infty^1 \longrightarrow X_\infty \longrightarrow Y_\infty \longrightarrow 0, \qquad (4.15)$$

which we shall henceforth call the *fundamental exact sequence*.

Iwasawa's theorem (Theorem 4.4.1) gives a precise analytic description of the $\Lambda(G)$-module $U_\infty^1/C_\infty^1$ in terms of the $p$-adic zeta function, while the modules $E_\infty^1/C_\infty^1$ and $Y_\infty$ occuring in (4.15) measure the discrepancy between $U_\infty^1/C_\infty^1$ and $X_\infty$. In fact, the following proposition, first observed by Iwasawa, shows that the two middle terms in (4.15) are isomorphic whenever the class number of $F_0 = \mathbb{Q}(\mu_p)^+$ is prime to $p$, which covers all known numerical examples.

**Proposition 4.5.3.** *Assume that the class number of $F_0 = \mathbb{Q}(\mu_p)^+$ is prime to $p$. Then*

$$E_\infty^1/C_\infty^1 = Y_\infty = 0.$$

*Proof.* Put $\Gamma_n = \mathrm{Gal}(F_\infty/F_n)$. and let $\Gamma = \Gamma_0$. Since there is a unique prime of $F_\infty$ above $p$ and this prime is totally ramified, it is a well known general fact in Iwasawa theory (see Appendix and [Wa, Chapter 13, Proposition 13.22]) that

$$(Y_\infty)_{\Gamma_n} \simeq \mathrm{Gal}(L_n/F_n) \qquad (4.16)$$

for all $n \geq 0$; here if $W$ is any $\Gamma$-module, $(W)_{\Gamma_n}$ will denote its $\Gamma_n$-coinvariants, and as above $L_n$ denotes the $p$-Hilbert class field of $F_n$. Hence our hypothesis that $L_0 = F_0$ implies that $(Y_\infty)_{\Gamma_0} = 0$ and hence $Y_\infty = 0$ by Nakayama's lemma. We then conclude from (4.16) that the class number of $F_n$ is prime to $p$ for all $n \geq 0$. To complete the proof, we now show that $E_n^1 = C_n^1$ for all $n \geq 0$. Indeed, since the classical class number formula asserts that the index of $D_n$ in $V_n$ is the class number of $F_n$, we have an exact sequence

$$0 \longrightarrow D_n^1 \longrightarrow V_n^1 \longrightarrow W_n \longrightarrow 0,$$

where $W_n$ is a finite group whose order is prime to $p$ since the class number of $F_n$ is prime to $p$. Thus the natural map from $D_n^1 \otimes \mathbb{Z}_p$ to $V_n^1 \otimes \mathbb{Z}_p$ is an isomorphism. It now follows from the commutative diagram

$$D_n^1 \otimes \mathbb{Z}_p \longrightarrow V_n^1 \otimes \mathbb{Z}_p$$

$$\downarrow \qquad\qquad\qquad \downarrow$$

$$C_n^1 \longrightarrow E_n^1$$

in which the vertical arrows are clearly surjective, that the inclusion of $C_n^1$ in $E_n^1$ is in fact an equality for all $n \geq 0$. Hence $E_\infty^1 = C_\infty^1$ and the proof is complete.    $\square$

**Corollary 4.5.4.** *Assume that the class number of $F_0 = \mathbb{Q}(\mu_p)^+$ is prime to $p$. Then*

$$X_\infty \simeq \Lambda(G)/I(G)\zeta_p.$$

*In particular, the main conjecture is true.*

We refer the reader to the Appendix for a brief discussion of the general facts about finitely generated modules over $\Lambda(G)$ which will be used in the remainder of this chapter. In particular, we recall that a finitely generated $\Lambda(G)$-module is said to be *torsion* if it is annihilated by an element of $\Lambda(G)$, which is not a divisor of zero.

**Proposition 4.5.5.** *All $\Lambda(G)$-modules appearing in the fundamental exact sequence (4.15) are finitely generated and torsion.*

*Proof.* The finite generation of these modules is easy. Indeed, it is clear from Theorem 4.4.1 that $U_\infty^1/C_\infty^1$ is finitely generated, and even cyclic as a $\Lambda(G)$-module, whence it is plain that $E_\infty^1/C_\infty^1$ is also finitely generated. The finite generation of both $X_\infty$ and $Y_\infty$ over $\Lambda(G)$ are also special cases of very general results in Iwasawa theory (see [Wa, Chapter 13]).

The proof that these modules are torsion lies deeper. For the module $U_\infty^1/C_\infty^1$, this amounts, thanks to Iwasawa's theorem, to showing that $\zeta_p \theta^+(e, 1)$ is not a zero divisor in $\Lambda(G)$ (see the proof of Theorem 4.4.1). But this follows from the remark about non-zero divisors made in the proof of Lemma 4.2.2, and the fact that $\zeta(1 - k)$ is non-zero for all even integers $k \geq 2$. For the remaining modules, we note that it is a well known general fact in Iwasawa theory (see [Wa, Chapter 13, Theorem 13.31]) that the analogue of $X_\infty$ for the cyclotomic $\mathbb{Z}_p$-extension of any totally real field is a torsion module over the Iwasawa algebra of the Galois group of the $\mathbb{Z}_p$-extension. However, in our case, this assertion follows easily from the fact that $U_\infty^1/C_\infty^1$ is torsion and that $Y_\infty$ is torsion. The latter assertion follows from (4.16), and the well known fact that the finiteness of the $\Gamma_0$-invariants of a compact

$\Gamma_0$-module implies that it is torsion over the Iwasawa algebra of $\Gamma_0$. This completes the proof.    □

We can now state the main conjecture (we recall that it is actually a theorem, whose proof will be completed in the later chapters). If $N$ is any finitely generated torsion $\Lambda(G)$-module, the structure theory (see Appendix) shows that we have an exact sequence of $\Lambda(G)$-modules

$$0 \longrightarrow \bigoplus_{i=1}^{r} \frac{\Lambda(G)}{\Lambda(G)f_i} \longrightarrow N \longrightarrow Q \longrightarrow 0,$$

where $f_i$ $(i = 1, \cdots, r)$ is a non-zero divisor, and $Q$ is finite. Then the *G-characteristic ideal* of $N$, which we denote by $\mathrm{ch}_G(N)$, is defined to be the ideal of $\Lambda(G)$ generated by the product $f_1 \cdots f_r$.

**Theorem 4.5.6.** (Main Conjecture) *We have*

$$\mathrm{ch}_G(X_\infty) = I(G)\zeta_p.$$

The completion of the proof of the Main Conjecture will occupy the next two chapters. However, in view of Iwasawa's theorem, and the multiplicativity of the characteristic ideal in exact sequences (see the Appendix), we deduce the following proposition immediately from the exact sequence (4.15).

**Proposition 4.5.7.** *The main conjecture is true if and only if* $\mathrm{ch}_G(Y_\infty) = \mathrm{ch}_G(E_\infty^1/C_\infty^1)$.

We stress that this last proposition is of theoretical interest rather than practical importance, because in view of Proposition 4.5.3, it amounts to the trivial assertion that $\Lambda(G) = \Lambda(G)$ in all known numerical cases.

## 4.6 Another Property of Cyclotomic Units

The aim of this section is to establish a further important property of the module $U_\infty^1/C_\infty^1$. This will help us to prove some additional results about the global units of the fields $F_n$ which will be needed in the arguments involving Euler systems in Chapter 6.

Henceforth, we shall use a deep result about the fields $F_n$, $(n \geq 0)$, which is due to Brumer [Br] and which is a special case of Leopoldt's conjecture. We remind the reader that Leopoldt's conjecture is unknown for arbitrary finite extensions of $\mathbb{Q}$.

**Theorem 4.6.1.** *Let $n$ be an integer $\geq 0$, and let $M_n$ denote the maximal abelian p-extension of $F_n$ which is unramified outside $p$. Then $M_n$ is a finite extension of $F_\infty$.*

We remark that the isomorphism (4.14) easily shows that the Galois group $\mathrm{Gal}(M_n/F_\infty)$ is finite if and only if the $\mathbb{Z}_p$-rank of $E_n^1$ is $[F_n : \mathbb{Q}] - 1$, where as in Definition 4.5.1, $E_n^1$ denotes the closure in $U_n^1$ of the global units of $F_n$ which are $\equiv 1 \bmod \mathfrak{p}_n$. This latter assertion is the more familiar form of Leopoldt's conjecture. In addition, it can be shown (see the appendix of [C1]) that the order of $\mathrm{Gal}(M_n/F_\infty)$ is given by a simple analytic formula which is essentially the inverse of the p-adic valuation of the residue at $s = 1$ of the $p$-adic zeta function of $F_n$. However, as we shall not use this formula in the rest of the proof, we do not enter into further details here.

**Definition 4.6.2.** *For each $m \geq 0$, we define*

$$\mathrm{N}_\infty(U_m^1) = \bigcap_{n \geq m} \mathrm{N}_{n,m}(U_n^1), \quad \mathrm{N}_\infty(K_m^\times) = \bigcap_{n \geq m} \mathrm{N}_{n,m}(K_n^\times)$$

*to be the subgroups of universal norms respectively of $U_m^1$ and $K_m^\times$.*

It is an easy exercise from local class field theory (see for example, [Iw2, Proposition 9]) that $\mathrm{N}_\infty(U_m^1)$ is the submodule of $U_m^1$ consisting of all elements whose norm to $\mathbb{Q}_p$ is equal to 1. Clearly, the natural projection from $U_\infty^1$ to $U_n^1$ induces a surjection

$$\alpha_{n,U} : (U_\infty^1)_{\Gamma_n} \longrightarrow \mathrm{N}_\infty(U_n^1), \tag{4.17}$$

where, as earlier, $\Gamma_n = \mathrm{Gal}(K_\infty/K_n)$. Further, since every element of $C_n^1$ is a universal norm, the natural projection from $C_\infty^1$ on $C_n^1$ induces a surjection

$$\alpha_{n,C} : (C_\infty^1)_{\Gamma_n} \longrightarrow C_n^1, \tag{4.18}$$

**Theorem 4.6.3.** *For all integers $n \geq 0$, we have (i) $(U_\infty^1/C_\infty^1)^{\Gamma_n} = 0$, (ii) $(U_\infty^1/C_\infty^1)_{\Gamma_n} = \mathrm{N}_\infty(U_n^1)/C_n^1$, and (iii) The natural map*

$$(C_\infty^1)_{\Gamma_n} \longrightarrow (U_\infty^1)_{\Gamma_n}$$

*maps $\mathrm{Ker}(\alpha_{n,C})$ isomorphically to $\mathrm{Ker}(\alpha_{n,U})$, both being isomorphic to $\mathbb{Z}_p$ with the trivial action of $G$.*

*Proof.* The proof, which is essentially due to Iwasawa [Iw2, Proposition 11], hinges on the following elementary observation. Let $\Phi_\infty$ denote the maximal abelian $p$-extension of $K_\infty$, and $\Phi_n$ the maximal abelian $p$-extension of $K_n$. Again, $G$ acts on $\mathrm{Gal}(\Phi_\infty/K_\infty)$ in the usual manner by inner automorphisms (see the analogous discussion for the global case in the previous section). Since $\Phi_n$ is plainly the maximal abelian extension of $K_n$ contained in $\Phi_\infty$, we have (see Lemma 3 of the Appendix)

$$(\mathrm{Gal}(\Phi_\infty / K_\infty))_{\Gamma_n} = \mathrm{Gal}(\Phi_n/K_\infty). \tag{4.19}$$

We can interpret this equation via local class field theory as follows. If $W$ is an abelian group, recall that its $p$-adic completion $\widehat{W}$ is defined by

$$\widehat{W} = \varprojlim W/p^m W$$

where the projective limit is taken with respect to the natural maps. Consider the $p$-adic completion $\widehat{K_n^\times}$ of $K_n^\times$, and define

$$Z_\infty = \varprojlim \widehat{K_n^\times},$$

where this projective limit is taken with respect to the homomorphisms induced by the norm maps. We can also consider the $p$-adic completion $\widehat{\mathrm{N}_\infty(K_n^\times)}$ of $\mathrm{N}_\infty(K_n^\times)$. Then the Artin map of local class field theory gives canonical $\Lambda(G)$-isomorphisms

$$Z_\infty \simeq \mathrm{Gal}(\Phi_\infty/K_\infty), \quad \widehat{\mathrm{N}_\infty(K_n^\times)} \simeq \mathrm{Gal}(\Phi_n/K_\infty).$$

Thus (4.19) shows that the projection from $Z_\infty$ to $\widehat{\mathrm{N}_\infty(K_n^\times)}$ induces a natural isomorphism

$$(Z_\infty)_{\Gamma_n} = \widehat{\mathrm{N}_\infty(K_n^\times)}. \tag{4.20}$$

Put $\tau_n = \gamma^{p^n} - 1$, where $\gamma$ is any fixed topological generator of $\Gamma_0$. Then (4.20) is equivalent to the assertion that the kernel of the natural surjection

$$\mathrm{pr}_{n,Z} : Z_\infty \longrightarrow \widehat{\mathrm{N}_\infty(K_n^\times)} \tag{4.21}$$

is precisely $\tau_n Z_\infty$. Let $W_n$ be the subgroup of $F_n^\times$ which is generated by all conjugates of $\pm v_n$ where $v_n = \zeta_n^{-1/2} - \zeta_n^{1/2}$. Clearly, $W_n$ is contained in $\mathrm{N}_\infty(K_n^\times)$, and since $v_n$ is a local parameter of $K_n$, the order valuation at $\mathfrak{p}_n$ gives an exact sequence of $G$-modules

$$0 \longrightarrow D_n \longrightarrow W_n \longrightarrow \mathbb{Z} \longrightarrow 0,$$

where $D_n$ is as in Definition 4.3.1. Passing to the $p$-adic completion, and recalling that the index of $D_n^1$ in $D_n$ is prime to $p$, we obtain the exact sequence of $G$-modules

$$0 \longrightarrow D_n^1 \otimes \mathbb{Z}_p \longrightarrow \widehat{W_n} \longrightarrow \mathbb{Z}_p \longrightarrow 0. \tag{4.22}$$

Similarly, we have an exact sequence of $G$-modules

$$0 \longrightarrow \mathrm{N}_\infty(U_n^1) \longrightarrow \widehat{\mathrm{N}_\infty(K_n^\times)} \longrightarrow \mathbb{Z}_p \longrightarrow 0. \tag{4.23}$$

As Leopoldt's conjecture is valid for $F_n$, we can identify $D_n^1 \otimes \mathbb{Z}_p$ with its image $C_n^1$ inside $\mathrm{N}_\infty(U_n^1)$. Hence the natural map from $\widehat{W_n}$ to $\widehat{\mathrm{N}_\infty(K_n^\times)}$ is injective, and it follows from the exact sequences (4.22) and (4.23) that

$$\widehat{\mathrm{N}_\infty(K_n^\times)}/\widehat{W_n} = \mathrm{N}_\infty(U_n^1)/C_n^1, \quad Z_\infty/W_\infty = U_\infty^1/C_\infty^1, \tag{4.24}$$

where we have written $W_\infty = \varprojlim \widehat{W_n}$, the projective limit being taken with respect to the norm maps.

Since the natural projection from $W_\infty$ to $\widehat{W_n}$ is plainly surjective, it follows from (4.20) that the kernel of the surjection

$$Z_\infty \xrightarrow{\;\mathrm{pr}_{n,Z}\;} \widehat{\mathrm{N}_\infty(K_n^\times)} \longrightarrow \widehat{\mathrm{N}_\infty(K_n^\times)}/\widehat{W_n}$$

is precisely $W_\infty \tau_n Z_\infty$. As

$$Z_\infty/W_\infty \tau_n Z_\infty = (Z_\infty/W_\infty)_{\Gamma_n},$$

assertion (ii) of Theorem 4.6.3 follows from the above remark and (4.24). But $U_\infty^1/C_\infty^1$ is a torsion $\Lambda(G)$-module by Theorem 4.4.1 and the fact that $\theta^+(e,1)\zeta_p$ is not a zero divisor in $\Lambda(G)$. Moreover, (ii) implies that

$$(U_\infty^1/C_\infty^1)_{\Gamma_n}$$

is finite because the validity of Leopoldt's conjecture shows that $\mathrm{N}_\infty(U_n^1)$ and $C_n^1$ have the same $\mathbb{Z}_p$-rank (of course we are using here the fact that $D_n$ has finite index in $V_n$ by the analytic class number formula). Hence, by a basic property of torsion $\Lambda(G)$-modules (see Appendix), we conclude that

$$(U_\infty^1/C_\infty^1)^{\Gamma_n}$$

is finite. But Iwasawa's theorem shows that $U_\infty^1/C_\infty^1$ is an elementary $\Lambda(G)$-module, and so it has no non-zero finite $\Lambda(G)$-submodules (see Appendix), whence assertion (i) follows.

To establish assertion (iii), we note that, thanks to the validity of (i) and (ii), we have the following commutative diagram with exact rows, where the right vertical arrow is an isomorphism:

$$
\begin{array}{ccccccccc}
0 & \longrightarrow & (C_\infty^1)_{\Gamma_n} & \longrightarrow & (U_\infty^1)_{\Gamma_n} & \longrightarrow & (U_\infty^1/C_\infty^1)_{\Gamma_n} & \longrightarrow & 0 \\
 & & \alpha_{n,C} \downarrow & & \alpha_{n,U} \downarrow & & \wr \downarrow & & \\
0 & \longrightarrow & C_n^1 & \longrightarrow & \mathrm{N}_\infty(U_n^1) & \longrightarrow & \mathrm{N}_\infty(U_n^1)/C_n^1 & \longrightarrow & 0.
\end{array}
$$

This immediately proves that the inclusion map in the top row induces an equality

$$\mathrm{Ker}(\alpha_{n,U}) = \mathrm{Ker}(\alpha_{n,C}).$$

On the other hand, applying the snake lemma to the following commutative diagram with exact rows

$$
\begin{array}{ccccccccc}
0 & \longrightarrow & U_\infty^1 & \longrightarrow & Z_\infty & \longrightarrow & \mathbb{Z}_p & \longrightarrow & 0 \\
 & & \times \tau_n \downarrow & & \times \tau_n \downarrow & & \times \tau_n \downarrow & & \\
0 & \longrightarrow & U_\infty^1 & \longrightarrow & Z_\infty & \longrightarrow & \mathbb{Z}_p & \longrightarrow & 0,
\end{array}
$$

we obtain an exact sequence

$$\mathbb{Z}_p \longrightarrow (U_\infty^1)_{\Gamma_n} \longrightarrow (Z_\infty)_{\Gamma_n} \longrightarrow \mathbb{Z}_p \longrightarrow 0. \tag{4.25}$$

where it is understood that the action of $G$ on both copies of $\mathbb{Z}_p$ in this sequence is the trivial action. Using (4.20) and (4.23), it is clear that we can rewrite this as the exact sequence

$$\mathbb{Z}_p \longrightarrow (U_\infty^1)_{\Gamma_n} \xrightarrow{\alpha_{n,U}} \mathrm{N}_\infty(U_n^1) \longrightarrow 0.$$

Hence the proof of (iii) will be complete provided we can show that the map on the left is injective. But, since clearly $(\Lambda(G))_{\Gamma_n}$ is a free $\mathbb{Z}_p$-module of rank $(p-1)p^n/2$, it follows from (4.10) that $(U_\infty^1)_{\Gamma_n}$ is also a free $\mathbb{Z}_p$-module of rank $(p-1)p^n/2$. On the other hand, as we have remarked earlier, local class field theory shows that $\mathrm{N}_\infty(U_n^1)$ has $\mathbb{Z}_p$-rank $(p-1)p^n/2 - 1$. Hence the map on the left above has to be injective, and the proof of Theorem 4.6.3 is now complete.    $\square$

## 4.7 Global Units

The aim of this section and the next one is to establish results for the finite extensions $F_n$ of $\mathbb{Q}$ which will be needed in Chapter 6 to carry

out the proof of Proposition 4.5.7 using Euler systems. We begin with
the results for global units, which are more delicate to prove than the
ones for ideal class groups, and our proofs will mainly be inspired by
those in Iwasawa's celebrated paper [Iw2].

Recall the canonical isomorphism (4.10)

$$\mathcal{L}^1 : U_\infty^1 \simeq \Lambda(G)$$

derived from the exact sequence (3.11). It is essentially clear from the
definition of the cyclotomic units (see Lemma 4.3.4) that $C_\infty^1$ is cyclic
over $\Lambda(G)$. It is a remarkable fact, whose proof requires deeper results
about the arithmetic of $F_\infty$, that the same is true for the module $E_\infty^1$
defined by (4.13).

**Theorem 4.7.1.** *The ideal $\mathcal{L}^1(E_\infty^1)$ is principal in $\Lambda(G)$.*

To establish this theorem and a related one at finite levels, we shall use
without proof, the following result due to Iwasawa (see [Iw4, Theorem
18]). In fact, it is valid for the analogue of $X_\infty$ over the cyclotomic
$\mathbb{Z}_p$-extension of an arbitrary finite extension of $\mathbb{Q}$.

**Proposition 4.7.2.** *$X_\infty$ has no non-zero finite $\Lambda(\Gamma_0)$-submodule, where
$\Gamma_0 = \mathrm{Gal}(F_\infty/F_0)$.*

We can now prove Theorem 4.7.1. Since $\mathcal{L}^1$ is a $\Lambda(G)$-homomorphism,
it suffices to show that $E_\infty^1$ is isomorphic to $\Lambda(G)$. Since $E_\infty^1$ is a submodule of a module isomorphic to $\Lambda(G)$, its torsion submodule is clearly
zero. Further, as $U_\infty^1/E_\infty^1$ is $\Lambda(G)$-torsion (see for example, Proposition
4.5.5), and $U_\infty^1$ is isomorphic to $\Lambda(G)$, it is clear that $E_\infty^1$ has well-
defined $\Lambda(G)$-rank equal to 1 (see Appendix). Hence by Theorem 1 of
the Appendix, there is an exact sequence of $\Lambda(G)$-modules

$$0 \longrightarrow E_\infty^1 \longrightarrow \Lambda(G) \longrightarrow Q \longrightarrow 0$$

where $Q$ is finite. Thus we must show that $Q = 0$. But since $\Gamma_0 =
\mathrm{Gal}(F_\infty/F_0)$ is pro-$p$, and $Q$ is finite, it suffices by Nakayama's lemma,
to prove that

$$Q^{\Gamma_0} = 0. \tag{4.26}$$

Taking $\Gamma_0$-invariants of the above exact sequence, we see immediately
that $Q^{\Gamma_0}$ must be the $\mathbb{Z}_p$-torsion submodule of $(E_\infty^1)_{\Gamma_0}$, since $\Lambda(G)_{\Gamma_0} =
\mathbb{Z}_p[\mathrm{Gal}(F_0/\mathbb{Q})]$ is a free $\mathbb{Z}_p$-module. On the other hand, we have the
exact sequence

$$0 \longrightarrow E_\infty^1 \longrightarrow U_\infty^1 \longrightarrow \mathrm{Gal}(M_\infty/L_\infty) \longrightarrow 0$$

given by Theorem 4.5.2. Taking $\Gamma_0$-invariants of this sequence, we obtain the exact sequence

$$0 \longrightarrow \operatorname{Gal}(M_\infty/L_\infty)^{\Gamma_0} \longrightarrow (E_\infty^1)_{\Gamma_0} \longrightarrow (U_\infty^1)_{\Gamma_0},$$

and again the group on the right is isomorphic to $\mathbb{Z}_p[\operatorname{Gal}(F_0/\mathbb{Q})]$ since $U_\infty^1$ is isomorphic to $\Lambda(G)$. Thus $(E_\infty^1)_{\Gamma_0}$ will be a free $\mathbb{Z}_p$-module, thereby proving (4.26), provided we can show that

$$\operatorname{Gal}(M_\infty/L_\infty)^{\Gamma_0} = 0. \tag{4.27}$$

To prove this, it suffices to show that

$$X_\infty^{\Gamma_0} = 0, \tag{4.28}$$

Thus, since $X_\infty$ has no non-zero finite $\Lambda(\Gamma_0)$-submodule by Proposition 4.7.2, to establish (4.28), we only need show that $X_\infty^{\Gamma_0}$ is finite. But, as $X_\infty$ is a torsion $\Lambda(\Gamma_0)$-module, $X_\infty^{\Gamma_0}$ is finite if and only if $(X_\infty)_{\Gamma_0}$ is finite (see Appendix). For any $n \geq 0$, recall that $M_n$ denotes the maximal abelian $p$-extension of $F_n$ which is unramified outside $p$. Clearly, $M_n$ is the maximal abelian $p$-extension of $F_n$ contained in $M_\infty$, whence it follows easily that (see Lemma 3 of the Appendix) that

$$(X_\infty)_{\Gamma_n} = \operatorname{Gal}(M_n/F_\infty). \tag{4.29}$$

Taking $n = 0$, we see that the validity of Leopoldt's conjecture for $F_0$ shows that $(X_\infty)_{\Gamma_0}$ is finite. This completes the proof of the theorem. □

To see what Theorem 4.7.1 implies for the fields $F_n$, we must introduce the corresponding universal norm subgroups.

**Definition 4.7.3.** *We define*

$$\mathrm{N}_\infty(V_m) = \bigcap_{n \geq m} \mathrm{N}_{n,m}(V_n), \quad \mathrm{N}_\infty(E_m^1) = \bigcap_{n \geq m} \mathrm{N}_{n,m}(E_n^1).$$

It is clear that $\mathrm{N}_\infty(E_m^1) = \overline{\mathrm{N}_\infty(V_m^1)}$. The following result is essentially due to Iwasawa [Iw2, Proposition 8], although our method of proof is different.

**Theorem 4.7.4.** *For each $n \geq 0$, let $R_n = \mathbb{Z}_p[\operatorname{Gal}(F_n/\mathbb{Q})]$. Then there is an $R_n$-isomorphism*

$$\mathrm{N}_\infty(E_n^1) \simeq R_n/\mathfrak{j}_n,$$

*where $\mathfrak{j}_n$ is isomorphic to $\mathbb{Z}_p$ with the trivial action of $\operatorname{Gal}(F_n/\mathbb{Q})$.*

*Proof.* We note first that by Theorem 4.7.1, $E_\infty^1$ is isomorphic to $\Lambda(G)$. Hence $(E_\infty^1)_{\Gamma_n}$ is isomorphic to $R_n$. Let

$$\alpha_{n,E} : (E_\infty^1)_{\Gamma_n} \longrightarrow \mathrm{N}_\infty(E_n^1)$$

be the surjection arising from the natural projection of $E_\infty^1$ on $E_n^1$. Thus, we must show that the kernel of $\alpha_{n,E}$ is isomorphic to $\mathbb{Z}_p$ with the trivial action of $G$. We have the commutative diagram in which the horizontal maps in the top row are induced from the inclusions $C_\infty^1 \subset E_\infty^1 \subset U_\infty^1$:

$$
\begin{array}{ccccc}
(C_\infty^1)_{\Gamma_n} & \longrightarrow & (E_\infty^1)_{\Gamma_n} & \longrightarrow & (U_\infty^1)_{\Gamma_n} \\
\downarrow {\scriptstyle \alpha_{n,C}} & & \downarrow {\scriptstyle \alpha_{n,E}} & & \downarrow {\scriptstyle \alpha_{n,U}} \\
C_n^1 & \longrightarrow & \mathrm{N}_\infty(E_n^1) & \longrightarrow & \mathrm{N}_\infty(U_n^1).
\end{array}
$$

By (iii) of Theorem 4.6.3, the composition of the two horizontal arrows in the top row is injective and induces an isomorphism from $\mathrm{Ker}(\alpha_{n,C})$ to $\mathrm{Ker}(\alpha_{n,U})$, both being isomorphic to $\mathbb{Z}_p$ with the trivial action. But Leopoldt's conjecture also implies that

$$(U_\infty^1 / E_\infty^1)^{\Gamma_n} = 0$$

(cf. the proof of Theorem 4.7.1). Hence the top horizontal map on the right in the above diagram is also injective. It follows therefore that the kernels of all three vertical arrows are isomorphic under the induced maps and equal to $\mathbb{Z}_p$ with $\mathrm{Gal}(F_n/\mathbb{Q})$ acting trivially. This completes the proof of the theorem.    □

**Proposition 4.7.5.** *For all $n \geq 0$, we have an $R_n$-isomorphism*

$$\left(E_\infty^1 / C_\infty^1\right)_{\Gamma_n} \simeq \mathrm{N}_\infty(E_n^1) / C_n^1. \tag{4.30}$$

*Proof.* Note first that $(E_\infty^1/C_\infty^1)^{\Gamma_n} = 0$ by (i) of Theorem 4.6.3. Hence, we have the following commutative diagram with exact rows

$$
\begin{array}{ccccccccc}
0 & \longrightarrow & (C_\infty^1)_{\Gamma_n} & \longrightarrow & (E_\infty^1)_{\Gamma_n} & \longrightarrow & (E_\infty^1/C_\infty^1)_{\Gamma_n} & \longrightarrow & 0 \\
& & \downarrow {\scriptstyle \alpha_{n,C}} & & \downarrow {\scriptstyle \alpha_{n,E}} & & \downarrow & & \\
0 & \longrightarrow & C_n^1 & \longrightarrow & \mathrm{N}_\infty(E_n^1) & \longrightarrow & \mathrm{N}_\infty(E_n^1)/C_n^1 & \longrightarrow & 0
\end{array}
$$
$$\tag{4.31}$$

where we recall that $\alpha_{n,C}$ and $\alpha_{n,E}$ are surjective. But it was shown at the end of the proof of Theorem 4.7.4 that the kernels of the first two vertical arrows coincide. Hence the last vertical arrow is an isomorphism and the proof of the proposition is complete.    □

**Theorem 4.7.6.** *For all $n \geq 0$, we have an $R_n$-isomorphism*

$$Y_\infty^{\Gamma_n} \simeq E_n^1/\mathrm{N}_\infty(E_n^1),$$

*where we recall that* $Y_\infty = \mathrm{Gal}(L_\infty/F_\infty)$. *In particular, the group on the right is finite and of order independent of $n$ when $n$ is sufficiently large.*

*Proof.* We first note that, since $\mathrm{Gal}(M_\infty/L_\infty)^{\Gamma_n} = 0$ by the validity of Leopoldt's conjecture, we have the commutative diagram of $R_n$-modules where the rows are exact,

$$
\begin{array}{ccccccccc}
0 & \longrightarrow & (E_\infty^1)_{\Gamma_n} & \longrightarrow & (U_\infty^1)_{\Gamma_n} & \longrightarrow & \mathrm{Gal}(M_\infty/L_\infty)_{\Gamma_n} & \longrightarrow & 0 \\
 & & \alpha_{n,E} \downarrow & & \alpha_{n,U} \downarrow & & \downarrow & & \\
0 & \longrightarrow & \mathrm{N}_\infty(E_n^1) & \longrightarrow & \mathrm{N}_\infty(U_n^1) & \longrightarrow & \mathrm{N}_\infty(U_n^1)/\mathrm{N}_\infty(E_n^1) & \longrightarrow & 0.
\end{array}
$$

$$(4.32)$$

The first two vertical arrows are surjective, and have equal kernels by the argument at the end of the proof of Theorem 4.7.4. Hence we conclude that we have an $R_n$-isomorphism

$$\mathrm{Gal}(M_\infty/L_\infty)_{\Gamma_n} \simeq \mathrm{N}_\infty(U_n^1)/\mathrm{N}_\infty(E_n^1). \qquad (4.33)$$

On the other hand, taking $\Gamma_n$-homology of the exact sequence

$$0 \longrightarrow \mathrm{Gal}(M_\infty/L_\infty) \longrightarrow X_\infty \longrightarrow Y_\infty \longrightarrow 0,$$

and recalling that $X_\infty^{\Gamma_n} = 0$, we obtain the exact sequence of $R_n$-modules

$$0 \longrightarrow Y_\infty^{\Gamma_n} \longrightarrow \mathrm{Gal}(M_\infty/L_\infty)_{\Gamma_n} \longrightarrow (X_\infty)_{\Gamma_n} \longrightarrow (Y_\infty)_{\Gamma_n} \longrightarrow 0. \quad (4.34)$$

As (see (4.29))

$$\mathrm{Gal}(M_n/F_\infty) = (X_\infty)_{\Gamma_n}, \quad \mathrm{Gal}(L_n F_\infty/F_\infty) = (Y_\infty)_{\Gamma_n}, \qquad (4.35)$$

it follows that we have the exact sequence

$$0 \longrightarrow Y_\infty^{\Gamma_n} \longrightarrow \mathrm{Gal}(M_\infty/L_\infty)_{\Gamma_n} \longrightarrow \mathrm{Gal}(M_n/L_n F_\infty) \longrightarrow 0. \qquad (4.36)$$

But the Artin map of global class field theory gives an $R_n$-isomorphism

$$\mathrm{Gal}(M_n/L_n F_\infty) \simeq \mathrm{N}_\infty(U_n^1)/E_n^1. \qquad (4.37)$$

Combining (4.33), (4.36), and (4.37), we conclude that

$$Y_\infty^{\Gamma_n} \simeq E_n^1/\mathrm{N}_\infty(E_n^1),$$

as required. But $Y_\infty^{\Gamma_n}$ is finite because $(Y_\infty)_{\Gamma_n} = \mathrm{Gal}(L_n/F_n)$ is finite. Hence $Y_\infty^{\Gamma_n}$ is contained in the maximal finite $\Lambda(\Gamma_0)$-submodule of $Y_\infty$ for all $n$, and it is, in fact equal to this module when $n$ is sufficiently large. Indeed, any finite $\Lambda(\Gamma_0)$-module is annihilated by a sufficiently large power of the maximal ideal of $\Lambda(\Gamma_0)$. This completes the proof of the theorem.    □

Theorem 4.7.1 shows that $\mathcal{L}^1(E_\infty^1) = \alpha\Lambda(G)$ for some $\alpha$ in $\Lambda(G)$. Since $E_\infty^1$ contains $C_\infty^1$, Theorem 4.4.1 proves that $\alpha$ must be a divisor of $\zeta_p\theta^+(e,1)$, say

$$\alpha\beta = \zeta_p\theta^+(e,1),$$

with $\beta$ in $\Lambda(G)$. But we have already remarked that $\zeta_p\theta^+(e,1)$ is not a divisor of zero in $\Lambda(G)$, whence the same is true for both $\alpha$ and $\beta$. We then clearly have a canonical isomorphism of $\Lambda(G)$-modules

$$\mathcal{J} : E_\infty^1/C_\infty^1 \simeq \Lambda(G)/\beta\Lambda(G). \tag{4.38}$$

We write $\mathrm{pr}_n : \Lambda(G) \longrightarrow R_n$ for the natural surjection.

**Theorem 4.7.7.** *For all $n \geq 0$, we have an $R_n$-isomorphism*

$$\mathcal{J}_n : \mathrm{N}_\infty(E_n^1)/C_n^1 \simeq R_n/\mathrm{pr}_n(\beta)R_n$$

*where $\beta$ is the element occuring in (4.38).*

*Proof.* Indeed, the isomorphism in (4.38) clearly induces an isomorphism of $R_n$-modules

$$\left(E_\infty^1/C_\infty^1\right)_{\Gamma_n} \simeq (\Lambda(G)/\beta\Lambda(G))_{\Gamma_n}. \tag{4.39}$$

Since $(\Lambda(G))_{\Gamma_n} = R_n$, we have the exact sequence

$$R_n \xrightarrow{\times\,\mathrm{pr}_n(\beta)} R_n \longrightarrow (\Lambda(G)/\beta\Lambda(G))_{\Gamma_n} \longrightarrow 0, \tag{4.40}$$

and hence

$$(\Lambda(G)/\beta\Lambda(G))_{\Gamma_n} = R_n/\mathrm{pr}_n(\beta)R_n,$$

thereby completing the proof of the theorem, granted Proposition 4.7.5.

Even though we shall not use it in the subsequent arguments, it may be worth pointing out a curious consequence of Proposition 4.7.5 and Theorem 4.7.6. Let $A_n$ denote the $p$-primary subgroup of the ideal class group of $F_n$. Combining the classical analytic class number formula with the validity of Leopoldt's conjecture for the field $F_n$, we have

$$\#(A_n) = \#(E_n^1/C_n^1) \qquad (n = 0, 1, \cdots).$$

Then we claim that $\#(A_n)$ is bounded as $n \longrightarrow \infty$ if and only if $\mathrm{N}_\infty(E_0^1) = C_0^1$. To prove this, put $P = E_\infty^1/C_\infty^1$. Assume first that $\mathrm{N}_\infty(E_0^1) = C_0^1$, whence by Proposition 4.7.5, $(P)_{\Gamma_0} = 0$, and so

$P = 0$ by Nakayama's lemma. Thus, again applying Proposition 4.7.5, it follows that $N_\infty(E_n^1) = C_n^1$ for all $n \geq 0$. We then conclude from the analytic class number formula above and Theorem 4.7.6 that $\#(A_n)$ is bounded as $n \longrightarrow \infty$. Conversely, assume that the cardinality of $A_n$ is bounded as $n \longrightarrow \infty$. By the analytic class number formula and Proposition 4.7.5, it follows that $\#((P)_{\Gamma_n})$ is bounded as $n \longrightarrow \infty$, whence we see easily from the structure theory that $P$ must be finite. But $P$ is a $\Lambda(G)$-submodule of $U_\infty^1/C_\infty^1$, and, as remarked earlier, Theorem 4.4.1 shows that $U_\infty^1/C_\infty^1$ has no non-zero finite $\Lambda(G)$-submodule. Hence $P = 0$, and so by Proposition 4.7.5, we have $N_\infty(E_0^1) = C_0^1$, as required. Finally, we point out that Greenberg [Gr] has indeed conjectured that $\#(A_n)$ is bounded as $n \longrightarrow \infty$. More generally, he makes the same conjecture for the cyclotomic $\mathbb{Z}_p$-extension of any totally real number field.

## 4.8 Ideal Class Groups

We end this chapter by establishing a rather weak result about the structure of the $p$-primary subgroup of the ideal class group of $F_m$ as a module over the group ring $R_m = \mathbb{Z}_p[\mathrm{Gal}(F_m/\mathbb{Q})]$. The method of proof is to use the structure theory for finitely generated torsion $\Lambda(G)$-modules (see Appendix), and to deduce the result for $F_m$ from this. Thus, as an inevitable consequence of the structure theory, there is a certain unknown finite $\Lambda(G)$-module appearing in the final result, which complicates the argument somewhat. However, it is interesting to note (see Theorem 4.8.2) that the same finite module occurs in the study of the quotient $E_n^1/N_\infty(E_n^1)$.

Recall that $L_\infty$ denotes the maximal abelian $p$-extension of $F_\infty$ which is everywhere unramified, and that $Y_\infty = \mathrm{Gal}(L_\infty/F_\infty)$, endowed with its natural structure as a $\Lambda(G)$-module. As explained in the proof of Proposition 4.5.5, $Y_\infty$ is a finitely generated torsion $\Lambda(G)$-module, and hence the structure theory (see Appendix) tells us that there is an exact sequence

$$0 \longrightarrow \bigoplus_{i=1}^{h} \frac{\Lambda(G)}{\Lambda(G)f_i} \longrightarrow Y_\infty \longrightarrow Q \longrightarrow 0, \qquad (4.41)$$

with $Q$ an unknown finite $\Lambda(G)$-module. Write $A_m$ for the $p$-primary subgroup of the ideal class group of $F_m$. Let $\mathrm{ann}(Q)$ be the annihilator ideal of $Q$ in $\Lambda(G)$.

**Theorem 4.8.1.** *For all sufficiently large $m$, there is an increasing filtration $\{\mathrm{Fil}^i(A_m) : i = 0, \cdots, h\}$ of $A_m$ by $R_m$-submodules with $\mathrm{Fil}^0(A_m) = 0$ and satisfying:-*

*(i) For $i = 1, \cdots, h$, we have an exact sequence of $R_m$-modules*

$$0 \longrightarrow Q_{i,m} \longrightarrow R_m / \operatorname{pr}_m(f_i)R_m \longrightarrow \operatorname{Fil}^i(A_m) / \operatorname{Fil}^{i-1}(A_m) \longrightarrow 0 \tag{4.42}$$

*where $Q_{i,m}$ is a finite $R_m$-module which is annihilated by $\operatorname{ann}(Q)$;*
*(ii) $A_m/ \operatorname{Fil}^h(A_m)$ is a finite $R_m$-module which is annihilated by $\operatorname{ann}(Q)$.*

Before beginning the proof, we note the following general algebraic facts about the category of all finitely generated torsion $\Lambda(G)$-modules. We refer the reader the Appendix for a more detailed discussion. Firstly, every finitely generated torsion $\Lambda(G)$-module has a maximal finite $\Lambda(G)$-submodule. Secondly, any elementary $\Lambda(G)$-module has no non-zero finite $\Lambda(G)$-submodule. Finally, given any finite $\Lambda(G)$-module $B$, the groups $\Gamma_m$ acts trivially on $B$ for all sufficiently large $m$.

Turning to the proof of the theorem, we let

$$W = \bigoplus_{i=1}^{h} \frac{\Lambda(G)}{\Lambda(G)f_i},$$

be the elementary module appearing in the exact sequence (4.41). Let $m \geq 0$ be any integer. Taking $\Gamma_m$-homology of the exact sequence (4.41), we obtain the long exact sequence

$$0 \longrightarrow W^{\Gamma_m} \longrightarrow Y_\infty{}^{\Gamma_m} \longrightarrow Q^{\Gamma_m} \longrightarrow W_{\Gamma_m} \xrightarrow{\varkappa_m} (Y_\infty)_{\Gamma_m} \longrightarrow Q_{\Gamma_m} \longrightarrow 0, \tag{4.43}$$

where $\varkappa_m$ is the natural map. But by (4.16) and class field theory, we have

$$(Y_\infty)_{\Gamma_m} = \operatorname{Gal}(L_m/F_m) = A_m.$$

Hence in particular, $(Y_\infty)_{\Gamma_m}$ is finite. This in turn implies that $(Y_\infty)^{\Gamma_m}$ is finite (see Appendix). Thus $W^{\Gamma_m}$ is a finite $\Lambda(G)$-submodule of the elementary module $W$, and so $W^{\Gamma_m} = 0$ by the remarks in the previous paragraph. Moreover, there exists an integer $m_0$ such that for all $m \geq m_0$, we have $Q^{\Gamma_m} = Q$ and $Y_\infty{}^{\Gamma_m} = Q'$ where $Q'$ is the maximal finite $\Lambda(G)$-submodule of $Y_\infty$. We assume from now on that $m \geq m_0$. But clearly (compare with (4.40)),

$$W_{\Gamma_m} = \bigoplus_{i=1}^{h} \frac{R_m}{\operatorname{pr}_m(f_i)R_m}. \tag{4.44}$$

Thus the sequence (4.43) can be rewritten as the exact sequence

$$0 \longrightarrow Q'' \longrightarrow \bigoplus_{i=1}^{h} \frac{R_m}{\operatorname{pr}_m(f_i)R_m} \xrightarrow{\varkappa_m} A_m \longrightarrow Q \longrightarrow 0,$$

where $Q'' = Q/Q'$. For $i = 1, \cdots, h$ we define $\mathrm{Fil}^i(A_m)$ to be the image in $A_m$ of the restriction of $\varkappa_m$ to the submodule of $W_{\Gamma_m}$ defined by

$$W_{i,m} := \bigoplus_{k=1}^{i} \frac{R_m}{\mathrm{pr}_m(f_k)R_m}.$$

Also, writing

$$Q''_{i,m} := Q'' \cap W_{i,m},$$

we then clearly have the exact sequence

$$0 \longrightarrow Q''_{i,m} \longrightarrow W_{i,m} \longrightarrow \mathrm{Fil}^i(A_m) \longrightarrow 0$$

for $i = 1, \cdots, h$. The theorem now follows on defining $\mathrm{Fil}^0(A_m) = 0$, $Q_{i,m} = Q''_{i,m}/Q''_{i-1,m}$ for $i = 1, \cdots, h$ and noting that $A_m/\mathrm{Fil}^h(A_m) = Q$.   $\square$

**Theorem 4.8.2.** *For all $m \geq 0$, the module $E_m^1/\mathrm{N}_\infty(E_m^1)$ is annihilated by $\mathrm{ann}(Q)$ where $Q$ is the finite module appearing in the exact sequence (4.41).*

*Proof.* This is immediate from the exact sequence (4.43), on recalling Theorem 4.7.6 and the fact that $W^{\Gamma_m} = 0$. This completes the proof.   $\square$

Finally, we note the following well-known lemma, which will also be used in the Euler system proof of Chapter 6.

**Lemma 4.8.3.** *For all $m \geq 0$, we have*

$$A_m^{\mathrm{Gal}(F_m/\mathbb{Q})} = 0.$$

*Proof.* Let $\varpi' = \mathrm{Gal}(F_0/\mathbb{Q})$ so that $\varpi'$ is a cyclic group of order $(p-1)/2$. Then we can identify $\mathrm{Gal}(F_m/\mathbb{Q})$ with the direct product of $\varpi'$ and the Galois group $\mathrm{Gal}(P_m/\mathbb{Q})$, where $P_m$ is the $m$-th layer of the cyclotomic $\mathbb{Z}_p$-extension of $\mathbb{Q}$. Since the degree of $F_m/P_m$ is prime to $p$, we see that the natural map induces an isomorphism

$$B_m \simeq A_m^{\varpi'},$$

where $B_m$ denotes the $p$-primary subgroup of the ideal class group of $P_m$. But, as is well-known $B_m = 0$ for all $m \geq 0$. To prove this last assertion, let $L'_\infty$ denote the maximal unramified abelian $p$-extension of

$$P_\infty = \bigcup_{n \geq 0} P_n.$$

Then, as $P_m/\mathbb{Q}$ is totally ramified at $p$, we have

$$\mathrm{Gal}(L'_\infty/P_\infty)_{\Gamma_m} = B_m,$$

where $\Gamma_m = \mathrm{Gal}(P_\infty/P_m)$. But clearly $B_0 = 0$ and so $L'_\infty = P_\infty$ by Nakayama's lemma. Hence $B_m = 0$ for all $m$ as asserted. This completes the proof of the lemma.   $\square$

# 5

# Euler Systems

## 5.1 Introduction

The aim of this chapter is to axiomatically define and study Euler systems for the tower $F_\infty$. This remarkable new method was discovered simultaneously and independently by Kolyvagin [Ko] and Thaine [Th], and thus, in comparison with most of the other basic tools used in the arithmetic of cyclotomic fields, is relatively recent. The notion of a general Euler system grew out of this work and has been extensively studied in, for example, [Ru2], [PR], [Ka]. We do not enter into a general discussion of Euler systems here, but work with the simplest notion needed for the proof of the main conjecture. In the final part of the chapter, we first establish the Factorization Theorem which goes back to Kolyvagin and Thaine, and then a variant of the Cebotarev Theorem due to Rubin. We stress that the Factorization Theorem is intuitively very surprising for the following reason. Fix an integer $m \geq 0$ and the field $F_m = \mathbb{Q}(\mu_{p^{m+1}})^+$. By an ingenious use of Kummer theory, it establishes relations in the ideal class group of $F_m$ by employing units (cyclotomic, or more generally the values of abstract Euler systems), which lie in tamely ramified cyclotomic extensions of $F_m$. It will then be shown in Chapter 6 that these new relations, when combined with Rubin's Cebotarev Theorem, enable one to prove that

$$\mathrm{ch}_G(Y_\infty) \text{ divides } \mathrm{ch}_G(E_\infty^1/C_\infty^1),$$

whence it is easy to complete the proof of the main conjecture.

## 5.2 Euler Systems

For motivation, we begin by introducing what is in fact, the only known concrete example of the abstract definition of an Euler system to be given at the end of this section.

Let $r \geq 2$ be an integer and let $a_1, \cdots, a_r$ be non-zero integers and $n_1, \cdots, n_r$ be integers with $\sum_{j=1}^{r} n_j = 0$.

**Definition 5.2.1.** *We define*

$$\alpha(T) = \prod_{j=1}^{r} (T^{-a_j/2} - T^{a_j/2})^{n_j}.$$

The alert reader will notice that this is just an avatar of the rational function studied in the earlier chapters, and which gives the Coleman power series of cyclotomic units. Formally, it should be viewed here as a rational function in the variable $T^{1/2}$.

**Definition 5.2.2.** *Let $S$ denote the finite set of primes consisting of 2 and all prime divisors of $a_1, \cdots, a_r$. We define the group*

$$W_S = \{\zeta \in \bar{\mathbb{Q}} : \zeta^m = 1 \text{ for some integer } m \geq 1 \text{ with } (m, S) = 1\}.$$

In other words, the group $W_S$ is the direct sum of all the $\mu_{q^\infty}$ with $q$ running over all primes not in $S$. Since 2 belongs to $S$, $W_S$ is uniquely divisible by 2, i.e. every $\zeta$ in $W_S$ has a unique square root lying in $W_S$, which we denote by $\zeta^{1/2}$. Using $\alpha(T)$, one can define a map

$$\phi_\alpha : W_S \longrightarrow \bar{\mathbb{Q}}^\times \tag{5.1}$$

by

$$\phi_\alpha(\zeta) = \alpha(\zeta) \text{ for } \zeta \neq 1, \quad \phi_\alpha(1) = \prod_{j=1}^{r} a_j^{n_j}.$$

Note that every cyclotomic unit of $F_n$ is of the form $\alpha(\zeta_n)$ for a primitive $p^{n+1}$-th root of unity $\zeta$, and a function $\alpha$ as above with the integers $a_1, \cdots, a_r$ prime to $p$. Hence we can view every cyclotomic unit of $F_n$ as giving rise to an Euler system.

**Lemma 5.2.3.** *For all $\zeta$ in $W_S$, the following assertions hold:-*

*(i) We have $\phi_\alpha(\zeta^{-1}) = \phi_\alpha(\zeta)$ and $\phi_\alpha(\zeta^\sigma) = \phi_\alpha(\zeta)^\sigma$ for all $\sigma \in \mathrm{Gal}(\bar{\mathbb{Q}}/\mathbb{Q})$;*
*(ii) If $q$ is a prime not in $S$, we have*

$$\prod_{\rho \in \mu_q} \phi_\alpha(\rho\zeta) = \phi_\alpha(\zeta^q);$$

*(iii) If $q$ is a prime not in $S$, then, provided $\zeta$ has order prime to $q$, we have the congruence*

$$\phi_\alpha(\rho\zeta) \equiv \phi_\alpha(\zeta) \mod \mathfrak{q}$$

*for all $\rho$ in $\mu_q$ and all primes $\mathfrak{q}$ lying over $q$.*

*Proof.* The first part of assertion (i) is clear and the rest follows on noting that by the uniqueness of square roots in $W_S$, we have for all $\sigma$ in $\mathrm{Gal}(\bar{\mathbb{Q}}/\mathbb{Q})$,

$$\sigma(\zeta^{1/2}) = (\sigma(\zeta))^{1/2}.$$

The proof of (ii) breaks up into two cases. Suppose first that $\zeta$ belongs to $\mu_q$. Then we must show that

$$\prod_{\substack{\rho \in \mu_q \\ \rho \neq 1}} \phi_\alpha(\rho) = 1.$$

But since the product of all elements in a cyclic group of odd order is equal to the identity, we have

$$\prod_{\substack{\rho \in \mu_q \\ \rho \neq 1}} \phi_\alpha(\rho) = \prod_{j=1}^{r} \left( \prod_{\substack{\rho \in \mu_q \\ \rho \neq 1}} (1 - \rho^{a_j}) \right)^{n_j}.$$

But, as $(q, a_j) = 1$ for $j = 1, \cdots r$, we see that the right hand side of this last equation is equal to $\prod_{j=1}^{r} q^{n_j} = 1$ since $\sum_{j=1}^{r} n_j = 0$. Suppose next that $\zeta$ does not belong to $\mu_q$. Again, using the above remark on cyclic groups of odd order, we see that

$$\prod_{\rho \in \mu_q} \phi_\alpha(\zeta\rho) = \prod_{j=1}^{r} \prod_{\rho \in \mu_q} (\zeta^{-a_j/2} - \rho^{a_j}\zeta^{a_j/2})^{n_j}.$$

But the right hand side of this expression is clearly equal to $\phi_\alpha(\zeta^q)$ completing the proof of (ii).

To establish (iii), first assume that $\zeta = 1$. Since $\alpha(1) = \phi_\alpha(1)$, and $\alpha(T)$ is a power series in $T - 1$ with coefficients in $\mathbb{Z}_q$, it is plain that $\phi_\alpha(\rho) \equiv \phi_\alpha(1) \bmod \mathfrak{q}$. Suppose now that $\zeta \neq 1$ has order prime to $q$. Then we have an expansion of the form

$$\phi_\alpha(\zeta T) = \sum_{n=0}^{\infty} c_n (T - 1)^n, \tag{5.2}$$

where the $c_n$ belong to $\mathbb{Z}_q[\zeta]$ for all $n \geq 0$. To prove this, we simply note that

$$(\zeta T)^{-a_j/2} - (\zeta T)^{a_j/2} = \sum_{n=0}^{\infty} d_n (T - 1)^n$$

where the $d_n$, $(n \geq 0)$ lie in $\mathbb{Z}_q[\zeta]$ and $d_0 = \zeta^{-a_j/2}(1 - \zeta^{a_j})$. But $d_0$ is a unit in the ring $\mathbb{Z}_q[\zeta]$ because $\zeta^{a_j}$ is a root of unity of order prime to $q$, and distinct from 1. Thus the above power series is a unit in the ring of formal power series in $T - 1$ with coefficients in $\mathbb{Z}_q[\zeta]$ and (5.2) follows. To finish the proof of (iii), we simply note that the expansion (5.2) converges for $T = \rho$ in $\mu_q$, and shows that $\phi_\alpha(\zeta\rho) - \phi_\alpha(\zeta)$ belongs to the proper ideal generated by $(\rho - 1)$ in $\mathbb{Z}_q[\zeta, \rho]$. This completes the proof of (iii). $\qquad\qquad\square$

In fact, we could continue with the proof of the main conjecture using only the functions $\phi_\alpha$ given in (5.1) in terms of $\alpha(T)$. However, the subsequent arguments will only use the properties (i)-(iii) of Lemma 5.2.3 and we therefore axiomatise the situation by making the following definition. Let $S$ be any finite set of prime numbers containing the prime 2. As earlier, take $W_S$ to be the set of all roots of unity in $\bar{\mathbb{Q}}^\times$ whose order is prime to $S$.

**Definition 5.2.4.** *An Euler system is a map* $\phi : W_S \longrightarrow \bar{\mathbb{Q}}^\times$ *such that the following axioms hold, where $\zeta$ denotes any element of $W_S$:-*
E1. *We have* $\phi(\zeta^{-1}) = \phi(\zeta)$ *and* $\phi(\zeta^\sigma) = \phi(\zeta)^\sigma$ *for all* $\sigma \in \mathrm{Gal}(\bar{\mathbb{Q}}/\mathbb{Q})$;
E2. *If $q$ is a prime not in $S$, we have*

$$\prod_{\rho \in \mu_q} \phi(\rho\zeta) = \phi(\zeta^q);$$

E3. *If $q$ is a prime not in $S$, then, provided $\zeta$ has order prime to $q$, we have the congruence*

$$\phi(\rho\zeta) \equiv \phi(\zeta) \bmod \mathfrak{q}$$

*for all $\rho$ in $\mu_q$ and all primes $\mathfrak{q}$ dividing $q$.*

We now study some basic properties of these Euler systems.

We shall need the following notation. For each odd integer $m \geq 1$, we put

$$\mathcal{H}_m = \mathbb{Q}(\mu_m), \quad H_m = \mathbb{Q}(\mu_m)^+. \tag{5.3}$$

For finite field extensions $L_1/L_2$, we write $\mathrm{N}_{L_1/L_2}$ for the norm map from $L_1$ to $L_2$. If $q$ is a prime with $(q, m) = 1$, we shall write $\mathrm{Fr}_q$ for the Frobenius element of $\mathrm{Gal}(\mathcal{H}_m/\mathbb{Q})$ and its restriction to $H_m$; i.e. $\mathrm{Fr}_q$ is the field automorphism which acts on $\mu_m$ by $\zeta \mapsto \zeta^q$.

For the rest of this chapter, $\phi$ will denote an arbitrary Euler system as defined above. Note that if $m$ is any integer prime to $S$, axiom E1 shows that $\phi(\zeta)$ belongs to $H_m$ for $\zeta \in \mu_m$.

**Lemma 5.2.5.** *Let $m \geq 1$ be any integer prime to $S$, and let $q$ be a prime number which does not divide $m$ and which does not lie in $S$. Then for all $\zeta$ in $\mu_m$ and all $\rho \neq 1$ in $\mu_q$, we have*

$$\mathrm{N}_{H_{mq}/H_m}\, \phi(\rho\zeta) = \frac{\phi(\zeta)^{\mathrm{Fr}_q}}{\phi(\zeta)}. \tag{5.4}$$

*Proof.* We first note that $m > 2$ since $m$ is prime to $S$. Hence $\mathrm{Gal}(\mathcal{H}_{mq}/\mathcal{H}_m)$ is isomorphic to $\mathrm{Gal}(H_{mq}/H_m)$, and both are of order $q-1$ since $(m, q) = 1$. In particular, these Galois groups act transitively on $\mu_q \setminus \{1\}$. Thus by Axiom E1,

$$\mathrm{N}_{H_{mq}/H_m}(\phi(\rho\zeta)) = \prod_{\substack{\eta \in \mu_q \\ \eta \neq 1}} \phi(\eta\zeta).$$

But, by axiom E2, the right hand side is equal to $\phi(\zeta^q)/\phi(\zeta)$. As

$$\phi(\zeta)^{\mathrm{Fr}_q} = \phi(\zeta^{\mathrm{Fr}_q}) = \phi(\zeta^q),$$

the proof of the lemma is complete. $\qquad\square$

**Lemma 5.2.6.** *Let $m \geq 1$ be any integer prime to $S$ Suppose that $q$ is a prime number which does not divide $m$ and does not lie in $S$. Let $n$ be any integer $\geq 1$. Then, for all $\zeta$ in $\mu_m$, and all primitive $q^{n+1}$-th roots of unity $\eta$, we have*

$$\mathrm{N}_{H_{mq^{n+1}}/H_{mq^n}}\, \phi(\eta\zeta) = \phi(\eta^q \zeta^{\mathrm{Fr}_q}).$$

*Proof.* Since $n \geq 1$, the group $\mathrm{Gal}(\mathcal{H}_{mq^{n+1}}/\mathcal{H}_{mq^n})$ is again isomorphic to $\mathrm{Gal}(H_{mq^{n+1}}/H_{mq^n})$, and both are of order $q$. Hence the orbit of $\eta$

under the action of these Galois groups consists of the set $\eta\rho$ such that $\rho$ is in $\mu_q$. Hence

$$\mathrm{N}_{H_{mq^{n+1}}/H_{mq^n}} \phi(\eta\zeta) = \prod_{\rho\in\mu_q} \phi(\eta\rho\zeta) = \phi(\eta^q\zeta^q) = \phi(\eta^q\zeta^{\mathrm{Fr}_q})$$

by axiom E2 and this completes the proof. □

For each $n \geq 0$, let $\eta_n$ denote a primitive $q^{n+1}$-th root of unity with $\eta_n^q = \eta_{n-1}$ for all $n \geq 1$. We immediately obtain the following corollary.

**Corollary 5.2.7.** *Under the same hypotheses as Lemma 5.2.6, define* $v_n = \phi(\eta_{n-1}\tau_{n-1})$, *where* $\tau_n = \mathrm{Fr}_q^{-n}(\zeta)$. *Then*

$$\mathrm{N}_{H_{mq^{n+1}}/H_{mq^n}} (v_{n+1}) = v_n \quad (n = 1, 2, \cdots).$$ □

## 5.3 Values of Euler Systems

The following theorem seems to suggest that Euler systems are not useful for studying ideal class groups for cyclotomic fields. Nevertheless, as we shall see later, an ingenious idea due to Kolyvagin and Thaine shows that this judgment is too hasty.

**Theorem 5.3.1.** *For any Euler system* $\phi : W_S \longrightarrow \bar{\mathbb{Q}}^\times$, *the value* $\phi(\eta)$ *is a unit in* $\mathbb{Q}(\eta)^+$ *for all* $\eta \neq 1$ *in* $W_S$.

We start with the following well known lemma. Let $\mathbb{Q}^{\mathrm{cyc}}$ be the unique Galois extension of $\mathbb{Q}$ with Galois group isomorphic to $\mathbb{Z}_q$, which is contained in the field $\mathbb{Q}(\mu_{q^\infty})$. If $L$ is an arbitrary finite extension of $\mathbb{Q}$, we define the cyclotomic $\mathbb{Z}_q$-extension $L^{\mathrm{cyc}}$ of $L$ to be the compositum $L\mathbb{Q}^{\mathrm{cyc}}$. For each $n \geq 0$, write $L_n$ for the unique extension of $L$ contained in $L^{\mathrm{cyc}}$ which is cyclic of degree $q^n$ over $L$.

**Lemma 5.3.2.** *Let* $q$ *be any prime number, and let* $L$ *be any finite extension of* $\mathbb{Q}$. *Let* $z$ *in* $L^\times$ *be a norm from* $L_n$ *for all* $n \geq 0$. *Then* $\mathrm{ord}_\mathfrak{r}(z) = 0$ *for all finite primes* $\mathfrak{r}$ *of* $L$ *which do not lie over* $q$.

*Proof.* By the theory of cyclotomic fields, there are only finitely many primes of $\mathbb{Q}^{\mathrm{cyc}}$ lying above each rational prime, and it follows easily that the same is then true for $L^{\mathrm{cyc}}$. Moreover. the only primes which ramify in the extension $L^{\mathrm{cyc}}/L$ are the primes dividing $q$. Let $\mathfrak{r}$ be a prime of $L$ which does not divide $q$, and fix a compatible system of primes

$\mathfrak{r}_n$ of $L_n$ above $\mathfrak{r}$. Since $\mathfrak{r}$ is unramified in $L^{cyc}$ and its decomposition group has finite index in the Galois group of $L^{cyc}$ over $L$, it follows that $f_n = [k_n : k] \to \infty$ as $n \to \infty$, where $k_n$ and $k$ are the respective residue fields of $\mathfrak{r}_n$ and $\mathfrak{r}$. But we have $\mathrm{N}_{L_n/L}(\mathfrak{r}_n) = \mathfrak{r}^{f_n}$. Thus the fact that $z$ is a norm from $L_n$ for every $n$ implies that $\mathrm{ord}_{\mathfrak{r}}(z)$ is divisible by $f_n$ for all $n \geq 1$, and so it follows that $\mathrm{ord}_{\mathfrak{r}}(z) = 0$. This completes the proof. $\qquad\square$

We can now prove the above theorem. Let $t$ be the exact order of $\eta$, so that $t \neq 1$. Take $q$ to be any prime dividing $t$, say $t = t_1 q^{m+1}$, where $m \geq 0$ and $t_1$ is prime to $q$. Clearly there exists a $\zeta$ in $\mu_{t_1}$ and a primitive $q^{m+1}$-th root of unity $\rho_m$ such that

$$\eta = \rho_m \cdot \mathrm{Fr}_q^{-m}(\zeta).$$

But by Corollary 5.2.7, $\phi(\eta)$ is a norm from $H_{t_1 q^{n+1}}$ for all $n \geq m$. Applying the above lemma with $L = H_{t_1 q^{n+1}}$, we conclude that any prime occurring in the factorization of $\phi(\eta)$ must divide $q$. Clearly if there is a second prime dividing $t$, we can carry out the same argument with this other prime and thereby deduce that no prime can occur in the factorization of $\phi(\eta)$, and so $\phi(\eta)$ is a unit. Thus we are left with the case in which $t_1 = 1$ and $t = q^{m+1}$. Now by Lemma 5.2.6. the norm from $H_{q^{m+1}}$ to $H_q$ of $\phi(\eta)$ is $\phi(\rho_0)$, where $\rho_0 = \rho_m^{q^m}$. Since the unique prime above $q$ is totally ramified in this extension, $\phi(\eta)$ is a unit if and only if $\phi(\rho_0)$ is a unit. But in fact $\phi(\rho_0)$ is a unit because, by axiom E2, we have

$$\mathrm{N}_{H_q/\mathbb{Q}}(\phi(\rho_0))^2 = \prod_{\substack{\zeta \in \mu_q \\ \zeta \neq \rho_0^{-1}}} \phi(\rho_0 \zeta) = 1.$$

This completes the proof of the theorem. $\qquad\square$

## 5.4 The Factorization Theorem

In this section, we discuss the ingenious idea of Kolyvagin and Thaine for using Euler systems to obtain relations in ideal class groups. The key to this is the Factorization Theorem, which is only discussed in the generality required for the proof of the main conjecture, rather than for an arbitrary abelian extension of $\mathbb{Q}$. With this in mind, we fix for the rest of this section the following data:-
(i) An odd prime number $p$
(ii) The field $F = F_m = \mathbb{Q}(\mu_{p^{m+1}})^+$ for some integer $m \geq 0$

(iii) An integer $t = p^a$ for some integer $a \geq m + 1$

(iv) A finite set $S$ of prime numbers such that always $2 \in S$ and $p \notin S$.

We shall also adopt the following notation. For each integer $r \geq 1$, we write

$$J_r = F(\mu_r)^+ = \mathbb{Q}(\mu_{p^{m+1}}, \mu_r)^+ \text{ and } \Delta_r = \text{Gal}(J_r/F). \qquad (5.5)$$

We write $I$ for the free abelian group on the non-zero prime ideals of $F$ written additively, and for each rational prime $q$, we write $I_q$ for the free abelian group on the primes of $F$ dividing $q$. Thus $I = \bigoplus_q I_q$. We define $Z_S$ to be the set of all square free positive integers which are prime to both $S$ and $p$. In addition, we write $Z_S^1$ for the subset of $Z_S$ consisting of all integers $n$ in $Z_S$ such that $n = q_1 \ldots q_k$ in $Z_S$ such that $q_i$ is a prime number with $q_i \equiv 1 \mod t$, for $i = 1, \cdots k$.

**Lemma 5.4.1.** *For all integers $n \geq 1$, the natural map*

$$F^\times/F^{\times^t} \longrightarrow (J_n^\times/J_n^{\times^t})^{\Delta_n} \qquad (5.6)$$

*is an isomorphism.*

*Proof.* By Kummer theory, the kernel and the cokernel of the map in (5.6) are $H^1(\Delta_n, \mu_t(J_n))$ and $H^2(\Delta_n, \mu_t(J_n))$, where $\mu_t(J_n) = \mu_t \cap J_n$. But $\mu_t(J_n) = 1$ since $J_n$ is totally real and $t$ is odd. $\quad\square$

**Lemma 5.4.2.** *Assume $n \in Z_S$. Then, for each prime $q$ dividing $n$, we have*

$$J_q \cap J_{n/q} = F, \quad J_n = J_q J_{n/q}.$$

*Moreover, each prime of $F$ dividing $q$ has ramification index equal to $q - 1$ in $J_n$.*

As an immediate corollary, we have the following:-

**Corollary 5.4.3.** *Assume $n \in Z_S$ is given by $n = q_1 \ldots q_k$, where the $q_i$ are distinct primes. Then*

$$\Delta_n \simeq \Delta_{q_1} \times \ldots \times \Delta_{q_k}.$$

We now prove the lemma. Put $k = p^{m+1} \cdot n/q$ so that $(k, q) = 1$. By the theory of cyclotomic fields, each prime of $F$ above $q$ is unramified in $J_{n/q}$, and totally ramified in $J_q$. Hence $J_q \cap J_{n/q} = F$. Also, we have that the compositum $J_q J_{n/q}$ must have degree $q - 1$ over $J_{n/q}$. But $[J_n : J_{n/q}] = q - 1$ and so $J_q J_{n/q} = J_n$. This completes the proof of the lemma. $\quad\square$

We now explain how we can operate on the values of Euler systems by canonical elements of the group rings $\mathbb{Z}[\Delta_r]$ to ensure that their images in $J_r^\times / J_r^{\times^t}$ are fixed by $\Delta_r = \mathrm{Gal}(J_r/F)$. By Lemma 5.4.1, this will then be the key to constructing elements in $F^\times / F^{\times^t}$ with interesting factorizations. Let $q$ be a prime number in $Z_S$. Recall that $\Delta_q = G(J_q/F)$ is a cyclic group of order $q - 1$. We now fix for the rest of this section a generator $\tau_q$ of $\Delta_q$, and define the elements

$$\mathcal{N}(q) = \sum_{\sigma \in \Delta_q} \sigma, \quad \mathcal{D}(q) = \sum_{k=0}^{q-2} k \tau_q^k, \tag{5.7}$$

of the integral group ring $\mathbb{Z}[\Delta_q]$, noting that $\mathcal{D}(q)$ depends on the choice of $\tau_q$. The proof of the following lemma is an evident calculation.

**Lemma 5.4.4.** *For each prime $q$ in $Z_S$, we have*

$$(\tau_q - 1)\mathcal{D}(q) = q - 1 - \mathcal{N}(q). \qquad \square$$

If $n = q_1 \dots q_k$ is a prime factorisation of $n$ in $Z_S$, we identify $\Delta_{q_i}$ with the subgroup of $\Delta_n$ given by $\mathrm{Gal}(\mathcal{H}_n/\mathcal{H}_{n/q_i})$. We then define the product

$$\mathcal{D}(n) = \mathcal{D}(q_1) \dots \mathcal{D}(q_k)$$

which we view as an element of $\Delta_n$. Now let

$$\phi : W_S \longrightarrow \bar{\mathbb{Q}}^\times \tag{5.8}$$

be an arbitrary Euler system. If $z$ is in $\mathcal{H}_n^\times$ and $\lambda$ is any element of $\mathbb{Z}[\Delta_n]$, $z^\lambda$ will denote $z$ acted on by $\lambda$.

**Proposition 5.4.5.** *Let $\rho$ be a primitive $p^{m+1}$-th root of unity. For each $n$ in $Z_S^1$, let $\xi_n$ be a primitive $n$-th root of unity. Then the class of $\phi(\rho\xi_n)^{\mathcal{D}(n)}$ in $J_n^\times / J_n^{\times^t}$ is fixed by $\Delta_n$.*

*Proof.* To lighten notation we define

$$\kappa_{n,q} = \mathcal{D}(n)(\tau_q - 1),$$

where $q$ denotes any prime in $Z_S^1$. The proof of the proposition is by induction on the number $k$ of prime factors of $n$, all of which are automatically distinct from $p$. Assume first that $n = q$ where $q$ is a prime with $q \equiv 1 \bmod t$. As a result of this latter congruence and Lemma 5.4.4, we have

$$\phi(\rho\xi_q)^{\kappa_{q,q}} = \phi(\rho\xi_q)^{-\mathcal{N}(q)} \bmod J_q^{\times^t}.$$

But, as $q$ is prime to $S$ and $p$, Lemma 5.2.5 shows that

$$\phi(\rho\xi_q)^{\mathcal{N}(q)} = \phi(\rho)^{\mathrm{Fr}_q-1}.$$

However, $\mathrm{Fr}_q = 1$ because $t = p^a$ where $a \geq m + 1$. This proves the propostion when $k = 1$.

Now assume that $k > 1$, and take $n = q_1 \ldots q_k$. We suppose the proposition is true for all elements of $Z_S^1$ with less than $k$ prime factors. We can plainly write $\xi_n = \xi_{q_1} \ldots \xi_{q_k}$, where $\xi_{q_i}$ is some primitive $q_i$-th root of unity. As $\tau_{q_1}, \cdots, \tau_{q_k}$ generate $\Delta_n$, it suffices to prove that

$$\phi(\rho\xi_n)^{\kappa_{n,q_i}} \in J_n^{\times^t}, \tag{5.9}$$

for all prime factors $q_i$ of $n$. Put

$$\mathcal{D}_i(n) = \prod_{\substack{j=1 \\ j \neq i}}^{k} \mathcal{D}(q_j).$$

By Lemma 5.4.4, and since $q_i \equiv 1 \bmod t$, we have

$$\phi(\rho\xi_n)^{\kappa_{n,q_i}} = \left(\phi(\rho\xi_n)^{\mathcal{D}_i(n)}\right)^{q_i-1-\mathcal{N}(q_i)} = \phi(\rho\xi_n)^{-\mathcal{D}_i(n)\mathcal{N}(q_i)} \bmod J_n^{\times^t}. \tag{5.10}$$

But applying Lemma 5.2.5 to the extension $J_n/J_{n/q_i}$, we have

$$\phi(\rho\xi_n)^{\mathcal{N}(q_i)} = \phi(\rho\xi_n/\xi_{q_i})^{\mathrm{Fr}_{q_i}-1}.$$

Substituting this equality in the right hand side of (5.10), and using our inductive hypothesis, we conclude that (5.9) is valid for $i = 1, \cdots, k$. This completes the proof of the proposition.  $\square$

The next lemma depends crucially on the fact that the primes above $q$ are tamely ramified in the extension $J_q/F$. Note that the map $l_q'$ of the lemma below depends on the choice of the generator $\tau_q$ of $\Delta_q$.

**Lemma 5.4.6.** *Assume that $q$ is a prime in $Z_S^1$. There is a natural homomorphism*

$$l_q' : (\mathcal{O}_F/q\mathcal{O}_F)^{\times} \longrightarrow I_q/tI_q, \tag{5.11}$$

*which commutes with the action of $\mathrm{Gal}(F/\mathbb{Q})$ and whose kernel is precisely the group of $t$-th powers in $(\mathcal{O}_F/q\mathcal{O}_F)^{\times}$.*

*Proof.* Since $q \in Z_S^1$, we have $q \equiv 1 \bmod t$. Further, as we have assumed that $t = p^a$ with $a \geq m + 1$, it follows that $q$ splits completely in $F$. Hence $I_q$ is a free $\mathbb{Z}[\mathrm{Gal}(F/\mathbb{Q})]$-module of rank 1. Now the extension $J_q/F$ has degree $q-1$, and each prime of $F$ above $q$ is totally and tamely ramified in this extension. Hence if $\mathfrak{q}$ denotes a prime of $F$ above $q$, the residue field of both $\mathfrak{q}$ and the unique prime of $J_q$ lying above $\mathfrak{q}$ is $\mathbb{F}_q$. Let $\pi_{\mathfrak{q}}$ denote some local parameter at the unique prime of $J_q$ above $\mathfrak{q}$. Since $\mathfrak{q}$ is totally and tamely ramified in $J_q$, elementary ramification theory shows that the map

$$\sigma \mapsto \pi_{\mathfrak{q}}/\sigma(\pi_{\mathfrak{q}})$$

denotes an isomorphism from $\Delta_q$ to $\mathbb{F}_q^{\times}$ which does not depend on the choice of the local parameter $\pi_{\mathfrak{q}}$. Thus our fixed generator $\tau_q$ of $\Delta_q$ maps under this isomorphism to a primitive root modulo $q$, which we denote by $\gamma_{\mathfrak{q}} = \pi_{\mathfrak{q}}^{1-\tau_q}$. We can now define the map $l_q'$. If $\alpha$ is an element of $\mathcal{O}_F$ with $(\alpha, q) = 1$, we have

$$\alpha \bmod \mathfrak{q} = \gamma_{\mathfrak{q}}^{a_{\mathfrak{q}}(\alpha)}, \text{ where } a_{\mathfrak{q}}(\alpha) \in \mathbb{Z}/(q-1)\mathbb{Z}.$$

We can then define

$$l_q'(\alpha \bmod q\mathcal{O}_F) = \sum_{\mathfrak{q}|q} (a_{\mathfrak{q}}(\alpha) \bmod t)\mathfrak{q}.$$

The right hand side makes sense since $q \equiv 1 \bmod t$ by hypothesis. One checks immediately that $l_q'$ has all the desired properties. □

We introduce the following notation. If $x$ belongs to $F^{\times}/F^{\times^t}$, we write

$$(x) = \sum_{\mathfrak{r}} \mathrm{ord}_{\mathfrak{r}}(x)\mathfrak{r} \bmod tI \qquad (5.12)$$

where $\mathfrak{r}$ runs over all finite primes of $F$, and

$$(x)_q = \sum_{\mathfrak{q}|q} \mathrm{ord}_{\mathfrak{q}}(x)\mathfrak{q} \bmod tI_q. \qquad (5.13)$$

**Definition 5.4.7.** *We define the subgroup $S_q$ of $F^{\times}/F^{\times^t}$ by*

$$S_q = \{x \in F^{\times}/F^{\times^t} : (x)_q = 0\}.$$

Suppose that $q$ is a prime such that $q \equiv 1 \bmod t$. We get a well-defined homomorphism

$$j_q : S_q \longrightarrow \mathcal{B}/\mathcal{B}^t, \text{ where } \mathcal{B} = (\mathcal{O}_F/q\mathcal{O}_F)^{\times}. \qquad (5.14)$$

by defining $j_q(x)$ to be the class in $\mathcal{B}$ of any integral representative of $x$ which has order valuation zero at all primes of $F$ dividing $q$. Thus finally we define the map

$$l_q : \mathcal{S}_q \longrightarrow I_q/tI_q \qquad (5.15)$$

to be the composite $l_q' \circ j_q$.

We now give one of the crucial definitions associated with Euler systems. From now on, fix a primitive $p^{m+1}$-th root of unity $\rho$. We shall suppress reference to $\rho$ in the notation when there is no danger of confusion. In the definition that follows, we are using of course Proposition 5.4.5. Recall that $Z_S^1$ is the set of all positive integers $n$ which are prime to $S$ and of the form $n = q_1 \ldots q_k$ such that the $q_i$ are distinct prime numbers with $q_i \equiv 1 \bmod t$ for $i = 1, \cdots, k$.

**Definition 5.4.8.** *Let $\phi$ be an Euler system as in (5.2.4). For each $n$ in $Z_S^1$, and each primitive $n$-th root of unity $\xi_n$, we define $\mathrm{R}_\phi(\xi_n)$ to be the unique element of $F^\times/F^{\times^t}$ such that its image under the isomorphism (5.6) is $\phi(\rho\xi_n)^{\mathcal{D}(n)} \bmod J_n^{\times^t}$.*

Note that, because $\phi(\rho\xi_n)^{\mathcal{D}(n)}$ is a unit in $J_n$, the ideal it generates is trivial. However, because there is ramification in the extension $J_n/F$, this does not imply that the ideal of $\mathrm{R}_\phi(\xi_n)$ in $I/I^t$ is zero. The next theorem, whose proof for the first time makes use of Axiom E3 for Euler systems, determines this ideal. Recall that if $x \in F^\times/F^{\times^t}$ and $q$ is any prime number, then $(x)_q$ given by (5.13) is its associated ideal in $I_q/tI_q$.

**Theorem 5.4.9.** *Let $\phi$ be any Euler system as in (5.2.4). Let $n = q_1 \ldots q_k$ be any element of $Z_S^1$, where the $q_i$ are prime numbers. Let $\xi_n$ be a primitive $n$-th root of unity and let $\xi_{q_i}$ be the component of $\xi_n$ in the group $\mu_{q_i}$ of $q_i$-th roots of unity. If $q$ is any prime number distinct from $q_1, \cdots, q_k$, then the ideal $(\mathrm{R}_\phi(\xi_n))_q = 0$. If $q = q_i$ for some $i = 1, \cdots, k$, then we have*

$$(\mathrm{R}_\phi(\xi_n))_{q_i} = l_{q_i}\left(\mathrm{R}_\phi(\frac{\xi_n}{\xi_{q_i}})\right), \qquad (5.16)$$

*where $l_{q_i}$ is the homomorphism given by (5.15).*

*Proof.* Since $\rho \neq 1$, Theorem 5.3.1 implies that $\phi(\rho\xi_n)^{\mathcal{D}(n)}$ is a unit in $J_n$. If $q$ is a prime distinct from $q_1, \cdots, q_r$, then each prime of $F$ above $q$ is unramified in the extension $J_n/F$, and hence $(\mathrm{R}_\phi(\xi_n))_q = 0$. Assume therefore that $q = q_i$ is one of the prime divisors of $n$. Then $q$ is unramified in the extension $J_{n/q}/\mathbb{Q}$ and we let $\mathrm{Fr}_q$ be the Frobenius element of $q$ in $\mathrm{Gal}(J_{n/q}/\mathbb{Q})$. Since $q \equiv 1 \bmod t$ and $t \geq m+1$, $\mathrm{Fr}_q$ belongs in fact to $\mathrm{Gal}(J_{n/q}/F)$. Let $z$ be any representative of $\mathrm{R}_\phi(\xi_n)$ in $F^\times$. Thus there exists $\beta$ in $J_n^\times$ such that

$$z = \frac{\phi(\rho\xi_n)^{\mathcal{D}(n)}}{\beta^t} \in F^\times. \tag{5.17}$$

To compute the $q$-part of the ideal of $z$, we write $\mathfrak{q}$ for any prime of $F$ above $q$, and choose some prime $\mathfrak{q}'$ of $J_n$ lying above $\mathfrak{q}$. Since $\mathfrak{q}'/\mathfrak{q}$ has ramification index $q-1$, and $\phi(\rho\xi_n)$ is a unit, it follows that

$$(\mathrm{R}_\phi(\xi_n))_q = \left( \sum_{\mathfrak{q}|q} \frac{t}{1-q} \, c_{\mathfrak{q}'} . \mathfrak{q} \right) \bmod t I_q, \tag{5.18}$$

where $c_{\mathfrak{q}'} = \mathrm{ord}_{\mathfrak{q}'}(\beta)$. Let $\pi_{\mathfrak{q}}$ denote a fixed local parameter at the unique prime of $J_q$ above $\mathfrak{q}$. Thus $\pi_{\mathfrak{q}}$ is a local parameter also at $\mathfrak{q}'$ since $J_n/J_q$ is unramified at the primes above $q$. Hence we can write

$$\beta = \pi_{\mathfrak{q}}^{c_{\mathfrak{q}'}} \alpha_{\mathfrak{q}},$$

where $\alpha_{\mathfrak{q}}$ is a unit at $\mathfrak{q}'$. Recall that $\tau_q$ is our fixed generator of $\mathrm{Gal}(J_n/J_{n/q})$. Since $\mathfrak{q}'$ is totally ramified in the extension $J_n/J_{n/q}$, we have

$$\alpha_{\mathfrak{q}}^{1-\tau_q} \equiv 1 \bmod \mathfrak{q}'.$$

Put $\gamma_{\mathfrak{q}} = \pi_{\mathfrak{q}}^{1-\tau_q}$. Thus we obtain

$$\beta^{1-\tau_q} \equiv \gamma_{\mathfrak{q}}^{c_{\mathfrak{q}'}} \bmod \mathfrak{q}'. \tag{5.19}$$

Hence $c_{\mathfrak{q}'}$ can be computed modulo $q-1$ as the $\mathfrak{q}$-component of $l'_q(\beta^{1-\tau_q})$ and this last expression can be determined as follows. Applying $(1-\tau_q)$ to (5.17), and noting the basic identity

$$(1 - \tau_q)\mathcal{D}(n) = (\mathcal{N}(q) + 1 - q)\mathcal{D}(n/q),$$

which follows from Lemma 5.4.4, we obtain

$$\beta^{(1-\tau_q)t} = \phi(\rho\xi_n)^{(\mathcal{N}(q)+1-q)\mathcal{D}(n/q)}. \tag{5.20}$$

But by Lemma 5.2.5,

$$\phi(\rho\xi_n)^{\mathcal{N}(q)} = \phi(\rho\xi_n/\xi_q)^{\mathrm{Fr}_q-1}. \tag{5.21}$$

On the other hand, we also know that there exists $\beta_q$ in $J_{n/q}$ such that

$$z_q = \frac{\phi(\rho\xi_n/\xi_q)^{\mathcal{D}(n/q)}}{\beta_q^t} \in F^\times,$$

and is prime to $q$. Thus $z_q$ is a representative of $R_\phi(\xi_n/\xi_q)$. Applying $\mathrm{Fr}_q - 1$ to this element and recalling that $\mathrm{Fr}_q$ fixes $F$, we obtain

$$\phi(\rho\xi_n/\xi_q)^{\mathcal{D}(n/q)(\mathrm{Fr}_q-1)} = \beta_q^{t(\mathrm{Fr}_q-1)}. \tag{5.22}$$

Hence substituting (5.22) and (5.21) into (5.20), it follows that

$$\beta_q^{t(\mathrm{Fr}_q-1)} = \beta^{t(1-\tau_q)}\phi(\rho\xi_n)^{(q-1)\mathcal{D}(n/q)}.$$

By the unqiueness of $t$-th roots in $J_n$, we conclude that

$$\beta_q^{\mathrm{Fr}_q-1} = \beta^{1-\tau_q}\phi(\rho\xi_n)^{\frac{q-1}{t}\mathcal{D}(n/q)}. \tag{5.23}$$

But by Axiom E3, we have that

$$\phi(\rho\xi_n) \equiv \phi(\rho\xi_n/\xi_q) \bmod \mathfrak{q}'.$$

Recalling that $\mathrm{Fr}_q$ acts on the residue field of $J_{n/q}$ at a prime above $q$ by raising to the $q$-th power, we conclude that

$$\beta^{1-\tau_q} \equiv \left(\frac{\beta_q^t}{\phi(\rho\xi_n/\xi_q)^{\mathcal{D}(n/q)}}\right)^{\frac{q-1}{t}} \equiv z_q^{\frac{1-q}{t}} \bmod \mathfrak{q}'.$$

Combining this congruence with (5.19), it follows that $c_{\mathfrak{q}'}$ is the $\mathfrak{q}$-component of

$$l'_q(\beta^{1-\tau_q}) = l'_q\left(z_q^{\frac{1-q}{t}}\right) = \left(\frac{1-q}{t}\right)l'_q(z_q) = \left(\frac{1-q}{t}\right)l_q(R_\phi(\xi_n/\xi_q)).$$

The assertion of the theorem is now clear from (5.18).    □

## 5.5 An Application of the Cebotarev Theorem

In this section, we follow Rubin [Ru3] and use the classical Cebotarev density theorem to establish a result which will play a central role in the inductive arguments with Euler systems given in the next chapter. We use the same fixed data, namely (i)-(iv), as in the previous section. In particular, we recall that $F = \mathbb{Q}(\mu_{p^{m+1}})^+$, and we define

$$A = p\text{-primary subgroup of the ideal class group of } F. \tag{5.24}$$

We also put

$$\Pi = \mathrm{Gal}(F/\mathbb{Q}). \tag{5.25}$$

**Theorem 5.5.1.** *Assume that we are given a class* $\mathfrak{c}$ *in* $A$, *a finite* $\Pi$*-submodule* $W$ *of* $F^\times/F^{\times^t}$, *and a* $\Pi$*-homomorphism*

$$\eta : W \longrightarrow (\mathbb{Z}/t\mathbb{Z})[\Pi].$$

*Then there exist infinitely many primes* $\mathfrak{q}$ *of* $F$, *say with* $\mathfrak{q}$ *lying above the rational prime* $q$, *such that (i)* $\mathfrak{q}$ *is in* $\mathfrak{c}$, *(ii)* $q \equiv 1 \bmod t$, *(iii)* $q$ *is not in* $S$, *(iv)* $W \subset S_q$, *and (v) there exists* $u \in (\mathbb{Z}/t\mathbb{Z})^\times$ *such that*

$$l_q(w) = u\eta(w)\,\mathfrak{q} \quad \textit{for all } w \in W, \tag{5.26}$$

*where* $l_q$ *is the homomorphism defined in* (5.15).

*Proof.* Let $L$ be the $p$-Hilbert class field of $F$, and put $\mathcal{F} = F(\mu_t)$. Then $\mathrm{Gal}(L/F)$ is isomorphic to $A$ as a $\Pi$-module by class field theory. In particular, we can view $\mathfrak{c}$ as an element of the Galois group $\mathrm{Gal}(L/F)$. Since $L/F$ is unramified, and $\mathcal{F}/F$ is totally ramified at the unique prime above $p$, we have

$$\mathcal{F} \cap L = F. \tag{5.27}$$

We next observe that

$$H^1(\mathrm{Gal}(\mathcal{F}/F), \mu_t) = 0. \tag{5.28}$$

Indeed, as $F$ is a real field, $H^0(\mathrm{Gal}(\mathcal{F}/F), \mu_t) = 0$, and so the Tate cohomology group $\widehat{H^0}(\mathrm{Gal}(\mathcal{F}/F), \mu_t) = 0$. But now (5.28) follows as $\mathrm{Gal}(\mathcal{F}/F)$ is cyclic, and the Herbrand quotient of its action on $\mu_t$ is 1. It follows from (5.28) that the natural map

$$F^\times/F^{\times^t} \longrightarrow \mathcal{F}^\times/\mathcal{F}^{\times^t}$$

is injective, and we identify $W$ with its image in the group on the right. We may therefore define

$$\mathcal{H} = \mathcal{F}(w^{1/t} : w \in W).$$

We claim that

$$\mathcal{H} \cap L = F. \tag{5.29}$$

Indeed, by Kummer theory, we have

$$\mathrm{Gal}(\mathcal{H}/\mathcal{F}) \simeq \mathrm{Hom}(W, \mu_t) \tag{5.30}$$

as $\Pi$-modules. Since the complex conjugation in $\Pi$ acts trivially on $W$ and on $\mu_t$ by $-1$, it follows that complex conjugation acts by $-1$ on $\mathrm{Gal}(\mathcal{H}/\mathcal{F})$. On the other hand,

$$\mathrm{Gal}(L\mathcal{F}/\mathcal{F}) = \mathrm{Gal}(L/F) = A$$

because $\mathcal{F}/F$ is totally ramified, and so complex conjugation acts like $+1$ on $\mathrm{Gal}(L\mathcal{F}/\mathcal{F})$ since $A$ is a subgroup of the ideal class group of the real field $F$. As $p$ is odd, it follows that

$$\mathcal{H} \cap L\mathcal{F} = \mathcal{F}.$$

Combining this last fact with (5.27), we conclude that (5.29) is valid.

Fix a primitive $t$-th root of unity $\zeta_t$ and define the $\mathbb{Z}/t\mathbb{Z}$-linear map

$$\iota : (\mathbb{Z}/t\mathbb{Z})[\Pi] \longrightarrow \mu_t$$

by $\iota(e) = \zeta_t$ where $e$ is the identity element of $\Pi$ and $\iota(g) = 1$ for all elements $g$ of $\Pi$ with $g \neq e$. Thus $\iota \circ \eta$ is a homomorphism from $W$ to $\mu_t$, and we define $\gamma$ to be the corresponding element of $\mathrm{Gal}(\mathcal{H}/\mathcal{F})$ under the isomorphism (5.30). Thus, by the definition of the Kummer isomorphism, we have

$$\iota \circ \eta(w) = \gamma(w^t)/w^{1/t} \text{ for all } w \in W. \tag{5.31}$$

In view of (5.29), we have

$$\mathrm{Gal}(\mathcal{H}L/F) = \mathrm{Gal}(L/F) \times \mathrm{Gal}(\mathcal{H}/F).$$

Hence there is a unique element $\sigma$ in $\mathrm{Gal}(\mathcal{H}L/F)$ which corresponds to the pair $(\mathfrak{c}, \gamma)$. By the Cebotarev density theorem, there exists infinitely many primes $\mathfrak{q}$ of $F$, which are of degree 1 and unramified in $F/\mathbb{Q}$ such that $\mathfrak{q}$ is unramified in $\mathcal{H}L$ and $\sigma$ belongs to the conjugacy class of the Frobenius elements of $\mathfrak{q}$ for the extension $\mathcal{H}L/F$. Writing $q$ for the rational prime below $\mathfrak{q}$, we now verify that $\mathfrak{q}$ satisfies all the assertions of the theorem, provided $q$ is sufficently large. Assertions (i), (iii), and (iv) are clear because $S$ and $W$ are finite by hypothesis. Assertion (ii) follows because the restriction of $\sigma$ to $\mathcal{F}$ is the identity and the Frobenius element of $q$ for the extension $\mathcal{F}/\mathbb{Q}$ acts on $\mu_t$ by raising to the $q$-th power. To prove (v), we consider the homomprhisms

$$f_i : W \longrightarrow \mathbb{Z}/t\mathbb{Z} \quad (i = 1, 2)$$

defined by

$$f_1(w) = \mathrm{ord}_{\mathfrak{q}}(l_q(w)), \quad f_2(w) = \mathrm{ord}_{\mathfrak{q}}(\eta(w)\mathfrak{q}).$$

We claim that

$$\mathrm{Ker}(f_1) = \mathrm{Ker}(f_2). \tag{5.32}$$

Clearly $w$ in $W$ belongs to $\mathrm{Ker}(f_1)$ if and only if $w$ is a $t$-th power modulo $\mathfrak{q}$. On the other hand, by the definition of the homomorphism

$\iota$, it is plain that $w$ belongs to $\mathrm{Ker}(f_2)$ if and only if $\iota \circ \eta(w) = 1$, which is equivalent to the assertion that $\sigma$ fixes every $t$-th root of $w$. But, as $\sigma$ is a Frobenius element for $\mathfrak{q}$, this is precisely the assertion that $w$ is a $t$-th power modulo $\mathfrak{q}$. Thus (5.32) holds, and so

$$W/\mathrm{Ker}(f_1) = W/\mathrm{Ker}(f_2)$$

is isomorphic to the same cyclic subgroup of $\mathbb{Z}/t\mathbb{Z}$ via the homomorphisms induced by $f_1$ and $f_2$, whence these two induced maps differ by multiplication by an element $u$ in $(\mathbb{Z}/t\mathbb{Z})^{\times}$. Therefore we have $f_1(w) = u f_2(w)$ for all $w$ in $W$. Since $l_q$ and $\eta$ are both $\Pi$-homomorphisms, it follows that

$$w \mapsto l_q(w) - u\eta(w)\mathfrak{q}$$

gives a $\Pi$-homomorphism from $W$ into $I_q$, whose image is contained in the subgroup of elements of $I_q$ whose $\mathfrak{q}$-component is zero. But the only $\Pi$-submodule of this latter subgroup is 0, and hence $l_q(w) = u\eta(w)\mathfrak{q}$, which is assertion (v). This completes the proof of the theorem.  □

We end by remarking that it is at first sight surprising that the above theorem holds for all $\Pi$-homomorphisms $\eta$ including the zero map since the homomorphism $l_q$ is never zero on the whole of $\mathcal{S}_q$. But there is no contradiction since assertion (v) holds only on the subgroup $W$ of $\mathcal{S}_q$.

# 6
# Main Conjecture

## 6.1 Introduction

The aim of this chapter is to complete the proof of the main conjecture using Euler systems. We broadly follow Rubin [Ru3], who showed how one could overcome considerable technical difficulties to use the ideas of Kolyvagin and Thaine to prove that

$$\mathrm{ch}_G(Y_\infty) \text{ divides } \mathrm{ch}_G(E^1_\infty/C^1_\infty)$$

in the fundamental exact sequence (4.15) of Chapter 4. Unlike the arguments of the earlier chapters which make essential use of the whole tower $F_\infty$, the Euler system argument takes place at a fixed finite extension $F$ of $\mathbb{Q}$ in $F_\infty$. The inductive argument then proceeds by a suitable choice of a sequence of degree one primes in $F$, and perhaps can be thought intuitively as some form of horizontal Iwasawa theory. This inductive argument is rather delicate to explain and we have based our exposition on that given in [C] for the analogous proof for elliptic curves with complex multiplication. To start the induction, it seems to be essential to know a precise statement about the Galois module structure of the universal norms in the unit groups of finite extensions $F/\mathbb{Q}$ in $F_\infty$, which goes back to Iwasawa (see for example, [Iw2, Proposition 8]). We end the chapter by presenting a well known counting argument based on the classical analytic class number formula which then shows that the above two characteristic ideals must coincide.

## 6.2 The Inductive Argument

As in the latter part of the previous chapter, we fix for the rest of this section the field

$$F = F_m = \mathbb{Q}(\mu_{p^{m+1}})^+$$

for some integer $m \geq 0$. We write $A = A_m$ for the $p$-primary subgroup of the ideal class group of $F$ and let

$$\Pi = \mathrm{Gal}(F/\mathbb{Q}), \quad R = R_m = \mathbb{Z}_p[\Pi].$$

Recall the exact sequence (4.41),

$$0 \longrightarrow \bigoplus_{i=1}^{h} \frac{\Lambda(G)}{\Lambda(G)f_i} \longrightarrow Y_\infty \longrightarrow Q \longrightarrow 0,$$

and also the isomorphism (4.38)

$$\mathcal{T} : E^1_\infty/C^1_\infty \simeq \Lambda(G)/\beta\Lambda(G),$$

where as always $G = \mathrm{Gal}(F_\infty/\mathbb{Q})$. Fix any annihilator $\delta$ in $\Lambda(G)$ of the finite module $Q$ above, but with the additional property that $R/\mathrm{pr}(\delta)R$ is finite; here, if $x$ is any element of $\Lambda(G)$, $\mathrm{pr}(x)$ denotes its image in $R$ under the natural map. In addition, as remarked after Theorem 4.7.7, $R/\mathrm{pr}(\beta)R$ is finite. Let $s$ be any fixed power of $p$ which annihilates both $R/\mathrm{pr}(\delta)R$ and $R/\mathrm{pr}(\beta)R$. We then define

$$t = \#(A)\#(Q)p^m s^{h+1} \tag{6.1}$$

where $h$ is the integer occurring in the exact sequence (4.41). We now define

$$\mathcal{R} = (\mathbb{Z}/t\mathbb{Z})[\Pi].$$

If $x$ is any element of $\Lambda(G)$, we write $x^\star$ for its image in $\mathcal{R}$ under the natural surjection from $\Lambda(G)$ onto $\mathcal{R}$. Finally, we fix a topological generator $\gamma$ of $G$. The goal of this section is to prove by induction the following divisibility assertion in $\mathcal{R}$.

**Theorem 6.2.1.** *For $i = 1, \cdots, h$, the product*

$$f_1^\star \cdots f_i^\star \text{ divides } ((\gamma - 1)\beta\delta^{i+1})^\star \tag{6.2}$$

*in $\mathcal{R}$.*

Before beginning the proof of the theorem, we explain how the Euler system used in it arises. Continuing to drop the subscript $m$ when it does not lead to confusion, we put

$$\mathrm{N}_\infty(V) = \mathrm{N}_\infty(V_m), \quad \mathrm{N}_\infty(E^1) = \mathrm{N}_\infty(E^1_m), \quad D = D_m,$$

reminding the reader that $D_m$ denotes the group of cyclotomic units of the field $F$. Recall that by Theorem 4.7.4, there is an isomorphism of $R$-modules

$$N_\infty(E^1) \simeq R/\mathfrak{j} \qquad (6.3)$$

where $\mathfrak{j} = \mathbb{Z}_p$ with the trivial action of $\Pi$. We stress that this isomorphism is fundamental for the inductive argument to be used in the proof of Theorem 6.2.1. Since the module on the left has no $p$-torsion, because $F$ is totally real, we conclude that

$$N_\infty(V) \, / \, N_\infty(V)^t = N_\infty(E^1) \, / \, N_\infty(E^1)^t \simeq \mathcal{R}/\mathfrak{n}, \qquad (6.4)$$

where $\mathfrak{n} = \mathbb{Z}/t\mathbb{Z}$ with the trivial action of $\Pi$. On the other hand, we recall that Theorem 4.7.7 gives an exact sequence of $R$-modules

$$0 \longrightarrow C^1 \longrightarrow N_\infty(E)^1 \longrightarrow R/\operatorname{pr}(\beta)R \longrightarrow 0.$$

Tensoring this sequence with $\mathbb{Z}/t\mathbb{Z}$ and noting that the index of $D^1$ (resp. $N_\infty(V)^1$) in $D$ (resp. in $N_\infty(V)$), is prime to $p$, we obtain the exact sequence

$$D/D^t \longrightarrow N_\infty(V)/N_\infty(V)^t \xrightarrow{\; j \;} \mathcal{R}/\beta^\star \mathcal{R} \longrightarrow 0 \qquad (6.5)$$

where $j$ is the induced map (of course, the map on the left need not be injective). We also have the exact sequence

$$0 \longrightarrow \Theta \longrightarrow N_\infty(V)/N_\infty(V)^t \longrightarrow F^\times/F^{\times^t}, \qquad (6.6)$$

for some finite group $\Theta$, which is induced by the inclusion of $N_\infty(V)$ in $F^\times$. Since the natural map

$$V/V^t \longrightarrow F^\times/F^{\times^t}$$

is clearly injective, it follows from the exact sequence

$$0 \longrightarrow N_\infty(V) \longrightarrow V \longrightarrow V/N_\infty(V) \longrightarrow 0$$

that

$$\Theta = (V/N_\infty(V))_t = (V^1/N_\infty(V^1))_t,$$

where, for any abelian group $M$, $M_t$ denotes the kernel of multiplication by $t$ on $M$. Hence it follows that $\Theta$ is annihilated by any element of $\mathcal{R}$ which kills $V^1/N_\infty(V)^1$. Therefore, by Theorem 4.8.2, $\Theta$ is annihilated by the element $\delta^\star$, where $\delta$ is our fixed non-zero annihilator of $Q$.

If $x$ is an element of $F^\times$, we write $\tilde{x}$ for its image in $F^\times/F^{\times^t}$. If, in addition, $x$ belongs to $N_\infty(V)$, we let $[x]$ denote its image in $N_\infty(V)/N_\infty(V)^t$. We must be careful to distinguish between the two because the map on the right in (6.6) is not injective. By virtue of the isomorphism (6.4) above, we can then make the following definition.

**Definition 6.2.2.** *Let $\varepsilon$ be any element of $N_\infty(V)$ such that $[\varepsilon]$ is mapped to the coset of $1 + \mathfrak{n}$ under the isomorphism (6.4).*

**Lemma 6.2.3.** *There exist integers $a_1, \cdots, a_r$ prime to $p$ and integers $n_1, \cdots, n_r$ with $\sum_{i=1}^{r} n_i = 0$, such that*

$$\beta^\star[\varepsilon] = [\alpha(\rho)]$$

*where*

$$\alpha(T) = \prod_{j=1}^{r} (T^{-a_j/2} - T^{a_j/2})^{n_j},$$

*and $\rho$ is a primitive $p^{m+1}$-th root of unity.*

*Proof.* Note that $\alpha(\rho)$ is in $D^1$ and also recall that any element in $D^1$ is of this form. By the exact sequence (6.5), the lemma is then clear since $\beta^\star j([\varepsilon]) = 0$.  □

To simplify notation, we write $\phi = \phi_\alpha$ (see (5.1)) for the Euler system corresponding to $\alpha(T)$ in the above lemma. In addition, if $x$ is any element of $R$, we also write $x^\star$ for its image under the natural surjection from $R$ onto $\mathcal{R}$.

**Lemma 6.2.4.** *Let $\lambda$ be any element of $R$ such that $\lambda\widetilde{\alpha(\rho)} = 1$ in $F^\times/F^{\times^t}$, where $\alpha(\rho)$ is the cyclotomic unit appearing in the previous lemma. Then we have $(\gamma^\star - 1)\delta^\star\beta^\star\lambda^\star = 0$ in $\mathcal{R}$.*

*Proof.* In view of the exact sequence (6.6), the hypothesis implies that

$$\lambda[\alpha(\rho)] \in \Theta.$$

Hence, as $\delta^\star$ annihilates $\Theta$, we have

$$\delta^\star\lambda[\alpha(\rho)] = 1.$$

By Lemma 6.2.3, it follows that

$$\beta^\star\delta^\star\lambda[\varepsilon] = 1.$$

By definition, $[\varepsilon]$ maps to the coset of $1 + \mathfrak{n}$ under the isomorphism (6.4). Thus

$$\beta^\star\delta^\star\lambda \in \mathfrak{n}$$

and so $(\gamma^\star - 1)\beta^\star\delta^\star\lambda = 0$ because $\Pi$ acts trivially on $\mathfrak{n}$. This completes the proof of the lemma.  □

We now begin the proof of Theorem 6.2.1 and first establish it for $i = 1$. Let $S$ be the set consisting of the prime 2 and all prime divisors of $a_1, \cdots, a_r$ where the $a_i$ are as in Lemma 6.2.3. Define

$$\mathcal{W}_0 = \mathcal{R}\nu_0, \text{ where } \nu_0 = \mathrm{R}_\phi(1) \in F^\times/F^{\times^t}, \tag{6.7}$$

where $\mathrm{R}_\phi$ is as in Definition 5.4.8; thus $\mathrm{R}_\phi(1) = \widetilde{\alpha(\rho)}$. Put

$$y_0 = \mathrm{pr}((\gamma - 1)\beta\delta) \in R. \tag{6.8}$$

In view of Lemma 6.2.4, we can define a $\Pi$-homomorphism

$$\eta_0 : \mathcal{W}_0 \longrightarrow \mathcal{R}$$

by $\eta_0(\lambda\nu_0) = \lambda y_0^\star$. As above, $A = A_m$ is the $p$-primary subgroup of the ideal class group of $F$. Choose $\mathfrak{c}_1$ to be any $R$-generator of $\mathrm{Fil}^1(A)$ in the filtration of $A$ given in Theorem 4.8.1. We now apply Theorem 5.5.1 with this choice of data. We conclude that there exists a prime $\mathfrak{q}_1$ of $F$ of degree 1, say with $\mathfrak{q}_1$ dividing $q_1$, such that (i) $\mathfrak{q}_1 \in \mathfrak{c}_1$, (ii) $q_1 \equiv 1 \bmod t$, (iii) $q_1 \notin S$, (iv) $\mathcal{W}_0 \subset \mathcal{S}_{q_1}$, and (v) there exists $u_0 \in (\mathbb{Z}/t\mathbb{Z})^\times$ such that

$$l_{\mathfrak{q}_1}(\nu_0) = u_0\eta_0(\nu_0)\mathfrak{q}_1 = u_0 y_0^\star \mathfrak{q}_1. \tag{6.9}$$

On the other hand, applying Theorem 5.4.9 with $n = q_1$, we obtain

$$l_{\mathfrak{q}_1}(\nu_0) = (\nu_1)_{\mathfrak{q}_1} = (\nu_1), \text{ where } \nu_1 = \mathrm{R}_\phi(\xi_1) \in F^\times/F^{\times^t},$$

with $\xi_1$ a primitive $q_1$-th root of unity. Thus we have

$$u_0 y_0^\star \mathfrak{q}_1 = (\nu_1) \text{ in } I/tI \tag{6.10}$$

whence

$$y_0\mathfrak{c}_1 = 0 \text{ in } A \tag{6.11}$$

because $tA = 0$. Note that we can immediately deduce the assertion of Theorem 6.2.1 for $i = 1$ from this last equation. Indeed, as $\mathrm{pr}(\delta)$ anihilates $Q$, it also annihilates $Q_1 = Q_{1,m}$ (see Theorem 4.8.1), and so it follows from (6.11) and the exact sequence (4.42) for $i = 1$, that $\mathrm{pr}(\delta)y_0$ annihilates $R/f_1R$, whence

$$(\delta y_0)^\star \in f_1^\star\mathcal{R}. \tag{6.12}$$

We now give in full detail the general inductive proof of Theorem 6.2.1. Let us fix classes $\mathfrak{c}_1, \cdots, \mathfrak{c}_h$ in $A$ such that $\mathfrak{c}_j$ belongs to $\mathrm{Fil}^j(A)$ and the quotient $\mathrm{Fil}^j(A)/\mathrm{Fil}^{j-1}(A)$ is generated over $R$ by the image of $\mathfrak{c}_j$, which we shall denote by $\tilde{\mathfrak{c}}_j$ for $j = 1, \cdots, h$. Suppose now that $i$ is any integer with $1 \leq i < h$. We assume by induction that we have already found degree one primes $\mathfrak{q}_1, \cdots, \mathfrak{q}_i$ of $F$ lying above distinct rational primes $q_1, \cdots, q_i$ respectively, which do not lie in $S$, and which have the following properties. Writing

$$\nu_j = R_\phi(\xi_{q_1} \cdots \xi_{q_j}) \text{ in } F^\times / F^{\times^t},$$

in the notation of Theorem 5.4.9, we then have :-

(i) $\mathfrak{q}_j$ is in $\mathfrak{c}_j$ for $j = 1, \cdots, i$

(ii) $q_j \equiv 1 \bmod t$ for $j = 1, \cdots, i$

(iii) There exist elements $y_{j-1}$ in $R$ such that there is an $\mathcal{R}$-homomorphism

$$\eta_{j-1} : W_{j-1} := \mathcal{R}\nu_{j-1} \longrightarrow \mathcal{R}$$

with $\eta_{j-1}(\nu_{j-1}) = y_{j-1}^\star$ for $j = 1, \cdots, i$

(iv) There exist $u_{j-1} \in (\mathbb{Z}/t\mathbb{Z})^\times$ such that

$$(\nu_j)_{q_j} = l_{q_j}(\nu_{j-1}) = u_{j-1}y_{j-1}^\star \mathfrak{q}_j, \text{ for } j = 1, \cdots, i$$

(v) $\delta^\star u_{j-1} y_{j-1}^\star = f_j^\star y_j^\star$ for $j = 1, \cdots, i - 1$.

We first note that assertions (i) to (v) imply the validity of Theorem 6.2.1 for $i$. Indeed, the equations (v) for $j = 1, \cdots, i - 1$ show that

$$(\delta^\star)^j y_0^\star \in f_1^\star \cdots f_j^\star y_j^\star \mathcal{R}. \tag{6.13}$$

On the other hand, assertions (i) and (iv) for $j = i$ imply that

$$y_{i-1}\tilde{\mathfrak{c}}_i = 0 \text{ in } \mathrm{Fil}^i(A) / \mathrm{Fil}^{i-1}(A).$$

Hence, using as before the exact sequence (4.42) for $i$, we conclude that

$$\mathrm{pr}(\delta)y_{i-1} \in f_i R. \tag{6.14}$$

Combining (6.13) and (6.14) for $j = i - 1$, it is clear that the conclusion of Theorem 6.2.1 is valid for $i$.

We now proceed to show that the above assertions (i) to (v) hold for $i + 1$. By the remarks above, this will establish Theorem 6.2.1 by induction. Our first step is to prove assertion (iii). Since the argument is rather delicate, we isolate the first step in the proof as a separate lemma.

**Lemma 6.2.5.** *Assume that the assertions (i) to (v) hold for an integer $i$ with $1 \leq i < h$. Let $\lambda$ be any element in $R$ with $\lambda\nu_i = 1$ in $F^\times / F^{\times^t}$. Then $\lambda.A = 0$.*

*Proof.* By virtue of the equation

$$l_{q_i}(\nu_{i-1}) = (\nu_i)_{q_i} = u_{i-1}y_{i-1}^{\star}q_i,$$

and our assumption that $\lambda\nu_i = 1$, we have

$$\lambda^{\star}y_{i-1}^{\star} = 0 \tag{6.15}$$

On the other hand, the equations in (v) for $j = 1, \cdots, i-1$ and the fact that $y_0 = \mathrm{pr}((\gamma - 1)\beta\delta)$ show that $y_{i-1}^{\star}$ divides $(\delta^i(\gamma - 1)\beta)^{\star}$ in $\mathfrak{R}$. But, by the definition of $s$, we have $\mathrm{pr}(\delta^i(\gamma - 1)\beta)$ divides $\mathrm{pr}(\gamma - 1)s^{i+1}$ in $R$. Moreover, by the definition of $t$, we see that $s^{i+1}$ divides $t/\#(A)$. Thus, by (6.15), we have

$$(t/\#(A))\lambda\,\mathrm{pr}(\gamma - 1) \in tR.$$

This implies that

$$\lambda\,\mathrm{pr}(\gamma - 1) \subset \#(A).R,$$

whence $\lambda\,\mathrm{pr}(\gamma - 1)A = 0$. But $\mathrm{pr}(\gamma - 1)$ is an automorphism of $A$, since $A^G = 0$ (see Lemma 4.8.3), and so we have proved the lemma. □

If $x$ is any element of $F^{\times}$, we write $(x)$ for its ideal in $I$, and define $\{x\}_{q_j} \in R$ by

$$(x)_{q_j} = \{x\}_{q_j}q_j;$$

here $(x)_{q_j}$ denotes the $q_j$-component of the ideal $(x)$. We now explain how to find a suitable $y_i$ in $R$ such that (iv) holds for $j = i + 1$. Pick a representative $b_i$ in $F^{\times}$ of $\nu_i$. By Theorem 5.4.9, the ideal $(b_i)$ can be written as

$$(b_i) = \sum_{j=1}^{i}\{b_i\}_{q_j}q_j + t\mathfrak{b}_i$$

where $\mathfrak{b}_i$ is prime to $q_1, \cdots, q_i$. Since the class of $q_j$ is $\mathfrak{c}_j$, $(1 \leq j \leq i)$, this equation shows that the class of $\mathfrak{b}_i$ belongs to $A$. Hence, as $tA = 0$, we obtain

$$\sum_{j=1}^{i}\{b_i\}_{q_j}\mathfrak{c}_j = 0.$$

In particular, recalling that $\tilde{\mathfrak{c}}_i$ denotes the image of $\mathfrak{c}_i$ in $\mathrm{Fil}^i(A)/\mathrm{Fil}^{i+1}(A)$, it follows from (i) of our inductive hypothesis that

$$\{b_i\}_{q_i}\tilde{\mathfrak{c}}_i = 0. \tag{6.16}$$

Since $\mathrm{pr}(\delta)$ annihilates the module $Q_{i,m}$, it follows from the exact sequence (4.42) that $\mathrm{pr}(\delta)\{b_i\}_{q_i}$ belongs to $\mathrm{pr}(f_i)R$, say

$$\mathrm{pr}(\delta)\{b_i\}_{q_i} = \mathrm{pr}(f_i)y_i, \quad (y_i \in R). \tag{6.17}$$

We fix any such $y_i$ and proceed to prove that

$$\lambda^\star y_i^\star = 0 \text{ in } \mathcal{R} \tag{6.18}$$

for all $\lambda$ in $R$ such that $\lambda\nu_i = 1$ in $F^\times/F^{\times^t}$. Fixing such a $\lambda$, it is clear that there exists $d_i$ in $F^\times$ such that

$$\lambda b_i = d_i^t. \tag{6.19}$$

We can write the ideal $(d_i)$ in the form

$$(d_i) = \sum_{j=1}^i \{d_i\}_{q_j} \mathfrak{q}_j + \mathfrak{d}_i, \tag{6.20}$$

where $\mathfrak{d}_i$ is prime to $\mathfrak{q}_1, \cdots, \mathfrak{q}_i$. It follows immediately from (6.19) that

$$\lambda\{b_i\}_{q_j} = t\{d_i\}_{q_j} \ (1 \le j \le i), \ \ \mathfrak{d}_i = \lambda\mathfrak{b}_i. \tag{6.21}$$

Noting that the class of $\mathfrak{b}_i$ belongs to $A$, the lemma above then shows that $\mathfrak{d}_i$ must be principal because $\mathfrak{d}_i = \lambda\mathfrak{b}_i$. Thus we obtain

$$\sum_{j=1}^i \{d_i\}_{q_j}\mathfrak{c}_j = 0, \tag{6.22}$$

and an entirely similar argument to the above then shows that there exists $z_i$ in $R$ such that

$$\mathrm{pr}(\delta)\{d_i\}_{q_i} = \mathrm{pr}(f_i)z_i. \tag{6.23}$$

As $\lambda\{b_i\}_{q_i} = t\{d_i\}_{q_i}$, we conclude from (6.17) and (6.23) that

$$\mathrm{pr}(f_i)z_i t = \mathrm{pr}(f_i)\lambda y_i. \tag{6.24}$$

But multiplication by $\mathrm{pr}(f_i)$ is injective on $R$ because $R/\mathrm{pr}(f_i)R$ is finite, and hence

$$z_i t = \lambda y_i \text{ in } R,$$

which clearly establishes (6.18). Thus the $\mathcal{R}$-homomorphism

$$\eta_i : W_i := \mathcal{R}\nu_i \longrightarrow \mathcal{R}$$

given by $\eta_i(\lambda\nu_i) = \lambda^\star y_i^\star$ is well-defined, and this establishes assertion (iii) for $i + 1$. Note also that assertion (v) for $j = i$ follows from the validity of (iv) for $j = i$, together with the fact that

$$(\nu_i)_{q_i} = \{b_i\}_{q_i}^\star \mathfrak{q}_i.$$

We now apply Theorem 5.5.1 to this homomorphism $\eta_i$ and the class $\mathfrak{c}_{i+1}$. We conclude that there exists a degree one prime $\mathfrak{q}_{i+1}$ of $F$ lying above a rational prime $q_{i+1}$ distinct from $q_1, \cdots, q_i$ and the elements of $S$ such that $\mathfrak{q}_{i+1} \in \mathfrak{c}_{i+1}$, $q_{i+1} \equiv 1 \bmod t$, and

$$l_{q_{i+1}}(\nu_i) = u_i y_i^\star \mathfrak{q}_{i+1}$$

for some $u_i$ in $(\mathbb{Z}/t\mathbb{Z})^\star$. By Theorem 5.4.9, we also have

$$(\nu_{i+1})_{q_i+1} = l_{q_i+1}(\nu_i)$$

and thus we have proven (iv) for $j = i + 1$. This completes the proof of the induction and hence assertions (i) to (v) hold for $i = 1, \cdots, h$ where we recall that $h$ is the number of direct summands appearing in the exact sequence (4.41). In particular, this establishes Theorem 6.2.1.                                              □

**Corollary 6.2.6.** $\mathrm{ch}_G(Y_\infty)$ *divides* $\mathrm{ch}_G(E_\infty^1/C_\infty^1)$.

*Proof.* We first claim that

$$f_1 \ldots f_h \text{ divides } (\gamma - 1)\beta\delta^{h+1} \text{ in } \Lambda(G). \tag{6.25}$$

Indeed, we have

$$\Lambda(G) = \varprojlim_m (\mathbb{Z}/p^m\mathbb{Z})[\mathrm{Gal}(F_m/\mathbb{Q})].$$

Since $t$ is a multiple of $p^{m+1}$, Theorem 6.2.1 for $i = h$ shows that the divisiblity assertion analogous to (6.25) holds in all of the group rings

$$(\mathbb{Z}/p^{m+1}\mathbb{Z})[\mathrm{Gal}(F_m/\mathbb{Q})]$$

for all $m \geq 0$, whence it clearly holds in $\Lambda(G)$ by a simple compactness argument.

The following argument shows that we can remove the power of $\delta$ in the assertion (6.25) above, where we recall that $\delta$ is any element of $\Lambda(G)$ which annihilates $Q$ and has the additional property that $R_m/\mathrm{pr}(\delta)R_m$ is finite for all $m \geq 0$. We recall that $\Lambda(G)$ is a direct product of $(p-1)/2$ copies of the unique factorization domain $\mathbb{Z}_p[[T]]$ and note that it therefore makes sense to define two elements of $\Lambda(G)$ to be relatively prime if each of their $(p-1)/2$ components are relatively prime. Since the module $Q$ appearing in (4.41) is finite, it is easy to see that we can find two relatively prime elements $\delta_1$ and $\delta_2$ which annihilate $Q$ and have the property that $R_m/\mathrm{pr}(\delta_1)R_m$ and $R_m/\mathrm{pr}(\delta_2)R_m$ are both finite for all $m \geq 0$. Indeed, $Q$ will be annihilated by the ideal of $\Lambda(G)$ given by taking the $k$-th power of the maximal ideal of $\mathbb{Z}_p[[T]]$ in each of the $(p-1)/2$ components for all sufficiently large positive integers $k$. One could then choose, for example, $\delta_1$ and $\delta_2$ to be the elements of $\Lambda(G)$, with $p^k$ and $T^k + p^k$, respectively, in each component. Thus it follows from (6.25) that

$$f_1 \ldots f_h \text{ divides } (\gamma - 1)\beta$$

in $\Lambda(G)$. However, we have already remarked that $(Y_\infty)_{\Gamma_0}$ is finite (see (4.16)), whence it follows that the product $f_1 \ldots f_h$ is relatively prime to $\gamma - 1$. Thus we conclude that

$$f_1 \ldots f_h \text{ divides } \beta$$

in $\Lambda(G)$ and the proof of the corollary is complete.          $\square$

## 6.3 Proof of the Main Conjecture

We can finally prove

**Theorem 6.3.1.** *We have* $\mathrm{ch}_G(Y_\infty) = \mathrm{ch}_G(E_\infty^1/C_\infty^1)$.

As explained in section 4.5 (see Proposition 4.5.7), this theorem, together with Iwasawa's theorem (Theorem 4.4.1) establishes the main conjecture at last.

We prove the theorem by invoking the classical analytic class number formula (see [H1, §11] or [Si]) for the field $F_0$. Indeed, recalling that $A_0$ denotes the $p$-primary subgroup of the ideal class group, $V_0$ denotes the group of units, and $D_0$ the group of cyclotomic units of $F_0$, the analytic class number formula asserts that

$$\#A_0 = \#((V_0 / D_0)(p)),$$

where $(V_0 / D_0)(p)$ denotes the $p$-primary subgroup of $V_0 / D_0$. But, since Leopoldt's conjecture is valid for $F_0$, we have

$$(V_0 / D_0)(p) = E_0^1 / C_0^1,$$

so that we can rewrite the class number formula as

$$\#(A_0) = \#(E_0^1 / C_0^1). \tag{6.26}$$

By (4.16) and global class field theory, we have

$$\#((Y_\infty)_{\Gamma_0}) = \#(A_0). \tag{6.27}$$

On the other hand, Proposition 4.7.5 shows that

$$\#(E_\infty^1 / C_\infty^1)_{\Gamma_0} = \#(\mathrm{N}_\infty(E_0^1)/C_0^1). \tag{6.28}$$

It follows from (6.26) and these last two formulae that

$$\#((Y_\infty)_{\Gamma_0})/\#(E_\infty^1/C_\infty^1)_{\Gamma_0} = \#(E_0^1/\mathrm{N}_\infty(E_0^1)). \tag{6.29}$$

But by Theorem 4.7.6, we have

$$\#(E_0^1/\mathrm{N}_\infty(E_0^1)) = \#(Y_\infty)^{\Gamma_0}.$$

Recalling that

$$(E_\infty^1/C_\infty^1)^{\Gamma_0} = 0$$

by (i) of Theorem 4.6.3, we have therefore shown that $Y_\infty$ and $E_\infty^1/C_\infty^1$ have finite $\Gamma_0$-Euler characteristics (see Appendix), and

$$\chi(\Gamma, Y_\infty) = \chi(\Gamma, E_\infty^1/C_\infty^1).$$

However, it is easily seen from Corollary 6.2.6 that $\mathrm{ch}_\Gamma(Y_\infty)$ divides $\mathrm{ch}_\Gamma(E_\infty^1/C_\infty^1)$. Hence by Corollary 2 of the Appendix, we conclude that we have

$$\mathrm{ch}_\Gamma(Y_\infty) = \mathrm{ch}_\Gamma(E_\infty^1/C_\infty^1),$$

whence again using Corollary 6.2.6, it follows that

$$\mathrm{ch}_G(Y_\infty) = \mathrm{ch}_G(E_\infty^1/C_\infty^1),$$

and this completes the proof of the main conjecture.    □

# 7

# Appendix

In the first part of this Appendix, we summarize for the convenience of the reader, the basic algebraic facts about modules over the Iwasawa algebras occurring in this book. In the latter parts, we recall several elementary lemmas from Iwasawa theory which are used in Chapters 4 and 6.

## A.1. Structure Theory

We recall that the Iwasawa algebra $\Lambda(\mathcal{G})$ of any profinite group $\mathcal{G}$ is defined by

$$\Lambda(\mathcal{G}) = \varprojlim \ \mathbb{Z}_p[\mathcal{G}/\mathcal{H}]$$

where $\mathcal{H}$ runs over all open normal subgroups of $\mathcal{G}$, and the inverse limit is taken with respect to the evident maps. It is endowed with the topology coming from the $p$-adic topology on the group rings of the finite quotients of $\mathcal{G}$. Modules over this Iwasawa algebra arise naturally in the following fashion. Let $M$ be any compact $\mathbb{Z}_p$-module on which $\mathcal{G}$-acts continuously on the left. Then

$$M = \varprojlim \ M_{\mathcal{H}}, \tag{A1}$$

where $M_{\mathcal{H}}$ denotes the largest quotient of $M$ on which $\mathcal{H}$ acts trivially. This is most easily seen by passing to the Pontrjagin dual of $M$, which we denote by $M^D$ and is defined to be the group of all continuous homomorphisms of $M$ into the discrete module $\mathbb{Q}_p/\mathbb{Z}_p$. We then note that

$$M^D = \bigcup (M^D)^{\mathcal{H}},$$

because $M^D$ is a discrete $\mathcal{G}$-module, and the previous assertion follows because $(M^D)^{\mathcal{H}}$ is dual to $M_{\mathcal{H}}$, and Pontrjagin duality changes inductive limits to projective limits. The left action of $\Lambda(\mathcal{G})$ on $M$ is evident from (A1).

The classical structure theory from commutative algebra (see [Bou, Chapter VII]), is usually only given for modules over the Iwasawa algebra of a group $\mathcal{G}$ which is isomorphic to $\mathbb{Z}_p$, or more generally $\mathbb{Z}_p^d$ for some integer $d \geq 1$. However, we shall be concerned with the case when $\mathcal{G}$ is either the Galois group over $\mathbb{Q}$ of the field generated by all $p$-power roots of unity, or its maximal real subfield, and it avoids unnecessary notational complexities to formulate the structure theory for such groups. Henceforth, we assume that $p$ is an odd prime number, and $\mathcal{G}$ is a group of the form $\pi \times \Gamma$, where $\pi$ is a quotient of $\mathbb{F}_p^{\times} = (\mathbb{Z}/p\mathbb{Z})^{\times}$ of order $k$, and $\Gamma$ is topologically isomorphic to the additive group of $\mathbb{Z}_p$. We write

$$\theta = \omega^{(p-1)/k}$$

where $\omega : \mathbb{F}_p^{\times} \longrightarrow \mathbb{Z}_p^{\times}$ is a Teichmüller character, i.e. $\omega(a) \equiv a \bmod p$. Thus the homomorphisms from $\pi$ to $\mathbb{Z}_p^{\times}$ are given precisely by the $\theta^i$, where $i$ runs over any complete set of residues modulo $k$. Hence, if $M$ is any $\Lambda(\mathcal{G})$-module, it will have the canonical decomposition

$$M = \bigoplus_{i \bmod k} M^{(i)}, \tag{A2}$$

where

$$M^{(i)} = e_{\theta^i} M, \quad e_{\theta^i} = \frac{1}{k}\sum_{\delta \in \pi} \theta^{-i}(\delta)\delta,$$

is the $\mathbb{Z}_p$-submodule of $M$ on which $\pi$ acts via $\theta^i$. As the elements of $\pi$ and $\Gamma$ commute, each $M^{(i)}$ is a $\Lambda(\Gamma)$-module. Applying this to $\Lambda(\mathcal{G})$ itself, it is clear that $\Lambda(\mathcal{G})^{(i)}$ is always a ring, and the next lemma describes it explicitly.

**Lemma 1** *Let $i$ be any integer modulo $k$. Then $\Lambda(\mathcal{G})^{(i)}$ is isomorphic to the ring $\Lambda(\Gamma)$, endowed with the action of $\pi$ via $\theta^i$.*

*Proof.* By definition, we have

$$\Lambda(\mathcal{G}) = \varprojlim \mathbb{Z}_p[\pi \times \Gamma/\Gamma_n],$$

where $\Gamma_n$ is the unique closed subgroup of $\Gamma$ of index $p^n$. Since $\mathbb{Z}_p[\pi \times \Gamma/\Gamma_n] = \mathbb{Z}_p[\pi][\Gamma/\Gamma_n]$, we conclude that

$$\Lambda(\mathcal{G})^{(i)} = \varprojlim \mathbb{Z}_p[\pi][\Gamma/\Gamma_n]^{(i)} = \varprojlim \mathbb{Z}_p[\Gamma/\Gamma_n],$$

where it is understood that $\pi$ acts on the group $\mathbb{Z}_p[\Gamma/\Gamma_n]$ via $\theta^i$. But

$$\varprojlim \mathbb{Z}_p[\Gamma/\Gamma_n] = \Lambda(\Gamma),$$

and so the proof of the lemma is complete. □

As a consequence of this lemma, we can essentially reduce many questions about a $\Lambda(\mathcal{G})$-module $M$ to the analogous questions for the $\Lambda(\Gamma)$-modules $M^{(i)}$ for $1 \le i \le k-1$, to which we can then apply the classical results of [Bou, Chapter VII]. For example, the module $M$ will be finitely generated over $\Lambda(\mathcal{G})$ if and only if each $M^{(i)}$ is finitely generated over $\Lambda(\Gamma)$. Also, defining the module $M$ to be $\Lambda(\mathcal{G})$-torsion if every element of $M$ is annihilated by a non-zero divisor in $\Lambda(\mathcal{G})$, it is clear that $M$ is $\Lambda(\mathcal{G})$-torsion if and only if each $M^{(i)}$ is $\Lambda(\Gamma)$-torsion, or equivalently $M$ itself is $\Lambda(\Gamma)$-torsion.

Let $M$ be a finitely generated $\Lambda(\mathcal{G})$-module. We shall say that $M$ has a well-defined $\Lambda(\mathcal{G})$-*rank* equal to $r$, if there is a $\Lambda(\mathcal{G})$-homomorphism from $M$ to $\Lambda(\mathcal{G})^r$ with $\Lambda(\mathcal{G})$-torsion kernel and cokernel. An equivalent definition is that each $M^{(i)}$, $(1 \le i \le k)$, should have $\Lambda(\Gamma)$-rank equal to $r$.

**Theorem 1.** *Let $M$ be a finitely generated $\Lambda(\mathcal{G})$-module. Assume (i) $M$ has a well-defined $\Lambda(\mathcal{G})$-rank equal to $r$, and (ii) the $\Lambda(\mathcal{G})$-torsion submodule of $M$ is zero. Then we have an exact sequence of $\Lambda(\mathcal{G})$-modules*

$$0 \longrightarrow M \longrightarrow \Lambda(\mathcal{G})^r \longrightarrow Q \longrightarrow 0,$$

*where $Q$ is a $\Lambda(\mathcal{G})$-module of finite cardinality.*

*Proof.* By assumptions (i) and (ii), we conclude from [Bou, Chapter VII] that we have an exact sequence of $\Lambda(\mathcal{G})$-modules

$$0 \longrightarrow M^{(i)} \longrightarrow \Lambda(\Gamma)^r \longrightarrow Q_i \longrightarrow 0$$

for all integers $i$ modulo $k$. Note that $r$ is independent of $i$ by assumption (i). Define the $\mathcal{G}$-module $Q$ by

$$Q = \bigoplus_{i \bmod k} Q_i$$

with the given action of $\Gamma$ on each summand, and with $\pi$ acting on $Q_i$ via the character $\theta^i$. Since by Lemma 1, we have the analogous decomposition

$$\Lambda(\mathcal{G})^r = \bigoplus_{i \bmod k} \Lambda(\Gamma)^r$$

the assertion of the theorem is plain. □

**Theorem 2.** *Let $M$ be a finitely generated torsion $\Lambda(\mathcal{G})$-module. Then there exists an exact sequence of $\Lambda(\mathcal{G})$-modules*

$$0 \longrightarrow \bigoplus_{j=1}^{r} \Lambda(\mathcal{G})/\Lambda(\mathcal{G})f_j \longrightarrow M \longrightarrow Q \longrightarrow 0.$$

*where $f_1, \cdots, f_r$ are non-zero divisors in $\Lambda(\mathcal{G})$, and $Q$ is $\Lambda(\mathcal{G})$-module of finite cardinality.*

*Proof.* For each integer $i \bmod k$ [Bou, Chapter VII] shows that we have an exact sequence of $\Lambda(\Gamma)$-modules

$$0 \longrightarrow \bigoplus_{j=1}^{r} \Lambda(\Gamma)/f_{j,i}\Lambda(\Gamma) \longrightarrow M^{(i)} \longrightarrow Q_i \longrightarrow 0,$$

where $f_{1,i}, \cdots, f_{r,i}$ are non-zero elements of $\Lambda(\Gamma)$, and $Q_i$ is a finite $\Lambda(\Gamma)$-module. Note that we can assume that $r$ is independent of $i$ simply by choosing some of the $f_{j,i}$'s to be one. Recalling that (cf. Lemma 1)

$$\Lambda(\mathcal{G}) = \bigoplus_{i \bmod k} \Lambda(\mathcal{G})^{(i)}, \text{ with } \Lambda(\mathcal{G})^{(i)} = \Lambda(\Gamma), \qquad (A3)$$

we define

$$f_j = \sum_{i \bmod k} f_{j,i}, \qquad (j = 1, \cdots, r).$$

Again taking

$$Q = \bigoplus_{i \bmod k} Q_i$$

with the given action of $\Gamma$ on each summand, and with $\pi$ acting on $Q_i$ via the character $\theta^i$, the assertion of the theorem follows. $\square$

Let $M$ be a finitely generated torsion $\Lambda(\mathcal{G})$-module. In view of Theorem 2, we may define the *characteristic ideal* $\mathrm{ch}_{\mathcal{G}}(M)$ by

$$\mathrm{ch}_{\mathcal{G}}(M) = f_1 \ldots f_m \Lambda(\mathcal{G}).$$

The uniqueness results in the structure theory (cf. [Bou, Chapter VII]) show that the ideal $\mathrm{ch}_{\mathcal{G}}(M)$ depends only on $M$, and not on the particular exact sequence in Theorem 2.

**Proposition 1.** *Let*

$$0 \longrightarrow M_1 \longrightarrow M_2 \longrightarrow M_3 \longrightarrow 0$$

*be an exact sequence of finitely generated torsion $\Lambda(\mathcal{G})$-modules. Then we have*

$$\mathrm{ch}_{\mathcal{G}}(M_2) = \mathrm{ch}_{\mathcal{G}}(M_1)\,\mathrm{ch}_{\mathcal{G}}(M_3).$$

*Proof.* It is proven in [Bou, Chapter VII] that the corresponding $\Lambda(\Gamma)$-characteristic ideals are multiplicative along exact sequences. Hence, as for each integer $i \bmod k$, we have an exact sequence of $\Lambda(\Gamma)$-modules

$$0 \longrightarrow M_1^{(i)} \longrightarrow M_2^{(i)} \longrightarrow M_3^{(i)} \longrightarrow 0,$$

the assertion of the propostion follows. □

**Lemma 2.** *Let $M$ be a finitely generated torsion $\Lambda(\mathcal{G})$-module, and suppose that $\mathrm{ch}_{\mathcal{G}}(M) = f_M \Lambda(\mathcal{G})$. Then, viewing $M$ as a $\Lambda(\Gamma)$-module via restriction of scalars, we have*

$$\mathrm{ch}_\Gamma(M) = f_{M,1} \ldots f_{M,k} \Lambda(\Gamma),$$

*where the $f_{M,i}$ are the components of $f_M$ in the decomposition (A3).*

*Proof.* This is immediate from the decomposition (A2) and the fact that $f_{M,i}$ is a generator of the $\Lambda(\Gamma)$-characteristic ideal of $M^{(i)}$ for $i = 1, \cdots, k$. □

Finally, a $\Lambda(\mathcal{G})$-module of the form

$$N = \bigoplus_{j=1}^{r} \Lambda(\mathcal{G})/\Lambda(\mathcal{G}) f_j$$

where $f_1, \cdots, f_r$ are non-zero divisors in $\Lambda(\mathcal{G})$ is called an *elementary* $\Lambda(\mathcal{G})$-module. A basic property of such elementary modules is that they have no non-zero finite $\Lambda(\mathcal{G})$-submodules. We omit the proof of this last assertion, simply noting that it follows easily on applying the Weierstrass preparation theorem to the $\Lambda(\Gamma)$-components of the elementary module.

## A.2. Γ-Euler Characteristics

As in the previous section, let $\Gamma$ be isomorphic to the additive group of $\mathbb{Z}_p$, and let $\Gamma_n$ denote the unique open subgroup of $\Gamma$ of index $p^n$. The augmentation homomorphism from $\Lambda(\Gamma)$ to $\mathbb{Z}_p$ induces an isomorphism $\Lambda(\Gamma)_\Gamma \simeq \mathbb{Z}_p$. If $g$ is any element of $\Lambda(\Gamma)$, we write $g(0)$ for its image under this isomorphism.

Let $M$ be a finitely generated torsion $\Lambda(\Gamma)$-module, and consider the homology groups $H_i(\Gamma, M)$ for $i \geq 0$. Since $\Gamma$ has $p$-homological dimension 1, we have

$$H_0(\Gamma, M) = (M)_\Gamma, \quad H_1(\Gamma, M) = M^\Gamma, \quad H_i(\Gamma, M) = 0 \quad (i \geq 2). \quad \text{(A4)}$$

We say that $M$ has finite $\Gamma$-Euler characteristic if the $H_i(\Gamma, M)$ $(i=0,1)$ are finite and, when they are finite, we define

$$\chi(\Gamma, M) = \#(H_0(\Gamma, M))/\#(H_1(\Gamma, M)).$$

Also, we write $g_M$ for any element of $\Lambda(\Gamma)$ such that

$$\mathrm{ch}_\Gamma(M) = g_M \Lambda(\Gamma).$$

**Proposition 2.** *Let $M$ be finitely generated torsion $\Lambda(\Gamma)$-module. Then the following assertions are equivalent:- (i) $H_0(\Gamma, M)$ is finite, (ii) $H_1(\Gamma, M)$ is finite and (iii) $g_M(0) \neq 0$. When these assertions hold, $\chi(\Gamma, M)$ is finite, and*

$$\chi(\Gamma, M) = |g_M(0)|_p^{-1}. \quad \text{(A5)}$$

*Proof.* By the structure theory of finitely generated $\Lambda(\Gamma)$-modules, we have an exact sequence of $\Lambda(\Gamma)$-modules

$$0 \longrightarrow \bigoplus_{j=1}^{m} \Lambda(\Gamma)/\Lambda(\Gamma)g_j \longrightarrow M \longrightarrow Q \longrightarrow 0, \quad \text{(A6)}$$

where $Q$ is finite, and $g_1, \cdots, g_m$ are non-zero elements of $\Lambda(\Gamma)$. Multiplying $g_1$ by a suitable unit, we can assume that $g_M = g_1 \ldots g_m$. Taking the long exact homology sequence of (A6), it is clear that it suffices to prove assertions (i), (ii) and (iii) are equivalent for each of the modules $R_i = \Lambda(\Gamma)/g_i\Lambda(\Gamma)$. But the short exact sequence

$$0 \longrightarrow \Lambda(\Gamma) \xrightarrow{\times g_i} \Lambda(\Gamma) \longrightarrow R_i \longrightarrow 0,$$

gives the long exact homology sequence

$$0 \longrightarrow H_1(\Gamma, R_i) \longrightarrow \mathbb{Z}_p \xrightarrow{\times g_i(0)} \mathbb{Z}_p \longrightarrow H_0(\Gamma, R_i) \longrightarrow 0.$$

It is now clear from this sequence that $H_0(\Gamma, R_i)$ is finite if and only if $H_1(\Gamma, R_i)$ is finite, and that both are equivalent to $g_i(0) \neq 0$. This proves the equivalence of assertions (i), (ii) and (iii). Moreover, writing $R = \bigoplus_{i=1}^{m} R_i$, it follows immediately from this exact sequence that if $g_M(0) \neq 0$, then $\chi(\Gamma, R)$ is finite, and $\chi(\Gamma, R) = |g_M(0)|_p^{-1}$. But $\chi(\Gamma, Q) = 1$ because $Q$ is finite. Hence by the multiplicativity of the Euler characteristic along short exact sequences, we conclude from (A6) that $\chi(\Gamma, M) = \chi(\Gamma, R)$ provided $g_M(0) \neq 0$. This completes the proof of the proposition. $\square$

**Corollary 1.** *Let $M$ be a finitely generated torsion $\Lambda(\Gamma)$-module, and let $\Gamma_n$ be the unique subgroup of $\Gamma$ of index $p^n$. Then for each integer $n \geq 0$, $(M)_{\Gamma_n}$ is finite if and only if $M^{\Gamma_n}$ is finite.*

*Proof.* Since $\Gamma_n$ has index $p^n$ in $\Gamma$, the ring $\Lambda(\Gamma)$ is a free $\Lambda(\Gamma_n)$-module of rank $p^n$. In particular $M$ is also a finitely generated $\Lambda(\Gamma_n)$-module. Recalling that both $\Lambda(\Gamma)$ and $\Lambda(\Gamma_n)$ are integral domains, it is also clear that $M$ has $\Lambda(\Gamma_n)$-rank zero because it has $\Lambda(\Gamma)$-rank zero. Hence the assertion follows on applying the above proposition with $\Gamma$ replaced by $\Gamma_n$. $\qquad\square$

**Corollary 2.** *Let $M_1$ and $M_2$ be two finitely generated torsion $\Lambda(\Gamma)$-modules such that (i) $\mathrm{ch}_\Gamma(M_1) \supset \mathrm{ch}_\Gamma(M_2)$, and (ii) $M_1$ and $M_2$ have finite $\Gamma$-Euler characteristics, with $\chi(\Gamma, M_1) = \chi(\Gamma, M_2)$. Then $\mathrm{ch}_\Gamma(M_1) = \mathrm{ch}_\Gamma(M_2)$.*

*Proof.* Let $g_{M_i}$ $(i = 1, 2)$ be a generator of $\mathrm{ch}_\Gamma(M_i)$. Then by (i), we have $g_{M_2} = g_{M_1} h$ for some $h$ in $\Lambda(\Gamma)$. But it follows from (ii) and the last assertion of the above proposition that $h(0)$ is a unit in $\mathbb{Z}_p$. Hence $h$ does not belong to the unique maximal ideal of the local ring $\Lambda(\Gamma)$, and therefore $h$ is a unit in $\Lambda(\Gamma)$. This completes the proof. $\qquad\square$

## A.3. Galois Groups and Iwasawa Theory

To help the reader, we briefly delve into the beginnings of Iwasawa theory, and explain in a little more detail the action of the Galois group of a $\mathbb{Z}_p$-extension on certain natural Iwasawa modules, which is used repeatedly in Chapter 4 and at the end of Chapter 6. Let $F$ be a field, and let $F_\infty$ be a $\mathbb{Z}_p$-extension of $F$, i.e. a Galois extension of $F$ whose Galois group is topologically isomorphic to the additive group of $\mathbb{Z}_p$. We write $\Gamma = \mathrm{Gal}(F_\infty/F)$ and, for each $n \geq 0$, we let $\Gamma_n$ denote the unique open subgroup of $\Gamma$ of index $p^n$. As usual, $F_n$ will denote the fixed field of $\Gamma_n$, so that

$$F_\infty = \bigcup_{n \geq 0} F_n.$$

Suppose now that we are given a Galois extension $\mathcal{M}_\infty$ of $F$ such that (i) $\mathcal{M}_\infty$ contains $F_\infty$, (ii) $\mathrm{Gal}(\mathcal{M}_\infty/F_\infty)$ is pro-$p$, and (iii) $\mathrm{Gal}(\mathcal{M}_\infty/F_\infty)$ is abelian. Let

$$X = \mathrm{Gal}(\mathcal{M}_\infty/F_\infty).$$

**Lemma 3.** *Under the above hypotheses, there is a natural action of* $\Gamma$ *on* $X$, *which extends to an action of the whole Iwasawa algebra. Moreover, if* $\mathcal{M}_n$ *denotes the maximal abelian extension of* $F_n$ *contained in* $F_\infty$, *we have*

$$\mathrm{Gal}(\mathcal{M}_\infty/\mathcal{M}_n) = \omega_n X, \quad (n \geq 0)$$

*where* $\omega_n = \gamma^{p^n} - 1$, *with* $\gamma$ *any fixed topological generator of* $\Gamma$.

*Proof.* For each $\sigma$ in $\Gamma$, let $\tilde{\sigma}$ be a lifting of $\sigma$ to $\mathrm{Gal}(\mathcal{M}_\infty/F)$. For $x$ in $X$, we then define

$$\sigma.x = \tilde{\sigma}x\tilde{\sigma}^{-1}.$$

The right hand side clearly only depends on $\sigma$ because $X$ is assumed to be abelian. One sees easily that this is an action of $\Gamma$, and that it is continuous when $X$ is endowed with the profinite topology. Since $X$ is a compact $\mathbb{Z}_p$-module, the remark made at the beginning of §A.1. shows that this action extends to an action of the whole Iwasawa algebra. To prove the final assertion of the lemma, put $\gamma_n = \gamma^{p^n}$ and let $h_n$ in $\mathrm{Gal}(\mathcal{M}_\infty/F_n)$ be a fixed lifting of $\gamma_n$. Since $\mathrm{Gal}(F_\infty/F_n)$ is topologically generated by $\gamma_n$, it is clear that every element of $\mathrm{Gal}(\mathcal{M}_\infty/F_n)$ is of the form $h_n^a x$ with $a$ in $\mathbb{Z}_p$ and $x$ in $X$. Since $\mathcal{M}_n$ is the maximal abelian extension of $F_n$ contained in $F_\infty$, the group $\mathrm{Gal}(\mathcal{M}_\infty/\mathcal{M}_n)$ is the closure of the commutator subgroup of $\mathrm{Gal}(\mathcal{M}_\infty/F_n)$, which we denote by $H_n$. We claim that

$$H_n = \omega_n X.$$

This follows because a simple commutator calculation shows that

$$[h_n^{a_1}x_1,\, h_n^{a_2}x_2] = \gamma_n^{a_2}(\gamma_n^{a_1} - 1)x_2 - \gamma_n^{a_1}(\gamma_n^{a_2} - 1)x_1,$$

for all $a_1$, $a_2$ in $\mathbb{Z}_p$, and $x_1$, $x_2$ in $X$. This completes the proof of the lemma.                                                        $\square$

In the applications given in Chapter 4, the field $F$ is the real subfield of $\mathbb{Q}(\mu_p)$ and $F_\infty$ is the real subfield of $\mathbb{Q}(\mu_{p^\infty})$. The field $\mathcal{M}_\infty$ is either the maximal abelian $p$-extension of $F_\infty$ which is unramified outside $p$ (denoted by $M_\infty$) or the maximal abelian $p$-extension of $F_\infty$ which is unramified everywhere (denoted by $L_\infty$). Writing $M_n$, (respectively $L_n$) for the maximal abelian $p$-extension of $F_n$ which is unramified outside of $p$ (resp. which is unramified everywhere), it is clear that $\mathcal{M}_n = M_n$ if $\mathcal{M}_\infty = M_\infty$. However, to prove that

$$\mathcal{M}_n = L_n F_\infty \text{ if } \mathcal{M}_\infty = L_\infty,$$

requires some additional arguments which are based on the fact that for our special $\mathbb{Z}_p$-extension $F_\infty/F$ of Chapter 4, there is a unique prime of $F$ above $p$ which is totally ramified in $F_\infty$. We omit the details, referring the reader to [Wa, §13.4].

# References

[BM]        S. Bentzen, I. Madsen, *Trace maps in algebraic K-theory and the Coates-Wiles homomorphism*, J. Reine Angew. Math. **411** (1990), 171–195.

[Bou]       N. Bourbaki, Elements of Mathematics, Commutative Algebra, Chapters 1-7, Springer, (1989).

[Br]        A. Brumer, *On the units of algebraic number fields*, Mathematika **14** (1967), 121–124.

[BCEMS]     J. Buhler, R. Crandall, R. Ernwall, T. Metsänkylä, M. Shokrollahi, *Irregular primes and cyclotomic invariants to 12 million*, J. Symbolic Comp. **31** (2001), 89–96.

[CW]        J. Coates, A. Wiles, *On the conjecture of Birch and Swinnerton-Dyer*, Invent. Math. **39** (1977), 223–251.

[CW2]       J. Coates, A. Wiles, *On p-adic L-functions and elliptic units*, J. Austral. Math. Soc. Ser. A **26** (1978), 1–25.

[C]         J. Coates, *Elliptic curves with complex multiplication and Iwasawa theory*, Bulletin London Math. Soc. **23** (1991), 321–350.

[C1]        J. Coates, *p-adic L-functions and Iwasawa's theory*, Algebraic Number Fields (Durham Symposium); ed. A. Frohlich, Academic Press, (1977) 269–353.

[CFKSV]     J. Coates, T. Fukaya, K. Kato, R. Sujatha, O. Venjakob, *The $GL_2$ main conjecture for elliptic curves without complex multiplication*, Publ. Math. IHES **101** (2005), 163–208.

[Co]        R. Coleman, *Division values in local fields*, Invent. Math. **53** (1979), 91–116.

[Co2]       R. Coleman, *Local units modulo circular units*, Proc. Amer. Math. Soc. **89** (1983), 1–7.

[Fe-W]      B. Ferrero, L. Washington, *The Iwasawa invariant $\mu_p$ vanishes for abelian number fields*, Ann. of Math. **109** (1979), 377–395.

[F]         J.-M. Fontaine, *Représentations p-adiques des corps locaux. I.*, in The Grothendieck Festschrift, Vol. II, Progr. Math. **87**, Birkhuser Boston, Boston, MA (1990), 249–309.

[Gr]        R. Greenberg, *On the Iwasawa invariants of totally real number fields*, Amer. J. Math. **98** (1976), 263–284.

[H]         H. Hasse, Bericht über neuere Untersuchungen und Probleme aus der Theorie der algebraischen Zahlkörper, Teil I: Klassenkörpertheorie. Teil Ia: Beweise zu Teil I. Teil II: Reziprozitätsgesetz. (German), Third ed. Physica-Verlag, Würzburg-Vienna (1970).

112    References

[H1]     H. Hasse, Über die Klassenzahl abelscher Zahlkörper, Akademie-Verlag Berlin, (1952).

[He]     J. Herbrand, *Sur les classes des corps circulaires*, J. Math Pures Appl. **11** (1932), 417–441.

[Iw1]    K. Iwasawa, *On the theory of cyclotomic fields*, Ann. of Math. **70** (1959), 530–561.

[Iw2]    K. Iwasawa, *On some modules in the theory of cyclotomic fields*, J. Math. Soc. Japan **20** (1964), 42–82.

[Iw3]    K. Iwasawa, *On p-adic L-functions*, Ann. of Math. **89** (1969), 198–205.

[Iw4]    K. Iwasawa, *On $\mathbb{Z}_l$-extensions of algebraic number fields*, Ann. of Math. **98** (1973), 246–326.

[Ka]     K. Kato, *Euler systems, Iwasawa theory and Selmer groups*, Kodai Math. J. **22** (1999), 313–372.

[Ka2]    K. Kato, *p-adic Hodge theory and values of zeta functions of modular forms*, Cohomologies p-adiques et applications arithmétiques. III, Astérisque **295** (2004), 117–290.

[Ko]     V. Kolyvagin, *Euler systems,* in The Grothendieck Festschrift, Vol. II, Progr. Math. **87**, Birkhuser Boston, Boston, MA (1990), 435–483.

[KL]     T. Kubota, H. Leopoldt, *Eine p-adische theorie der Zetawerte. I. Einführung der p-adischen Dirichletschen L-Funktionen*, Crelle J., **214/215** (1964), 328–339.

[L]      H. Leopoldt, *Eine p-adische theorie der Zetawerte. II. Die p-adische $\Gamma$-Transformation*, Crelle J., **274/275** (1975), 224–239.

[M]      K. Mahler, *An interpolation series for continuous functions of a p-adic variable*, Crelle J., **199** (1958), 23–34.

[MSD]    B. Mazur, P. Swinnerton-Dyer, *Arithmetic of Weil curves*, Invent. Math. **25** (1974), 1–61.

[MW]     B. Mazur, A. Wiles, *Class fields of abelian extensions of $\mathbb{Q}$*, Invent. Math. **76** (1984), 179–330.

[O]      R. Oliver, Whitehead groups of finite groups, London Mathematical Society Lecture Note Series **132**, Cambridge University Press, Cambridge, (1988).

[PR]     B. Perrin-Riou, *Systèmes d'Euler p-adiques et théorie d'Iwasawa*, Ann. Inst. Fourier **48** (1998), 1231–1307.

[Ri]     K.A. Ribet, *A modular construction of unramified p-extensions of $\mathbb{Q}(\mu_p)$*, Invent. Math. **34** (1976), 151–162.

[Ru]     K. Rubin, *The "main conjectures" of Iwasawa theory for imaginary quadratic fields*, Invent. Math. **103** (1991), 25–68.

[Ru2]    K. Rubin, Euler systems, Annals of Math. Studies **147**, Princeton University Press, Princeton, NJ, (2000).

[Ru3]    S. Lang, Cyclotomic fields I and II, Combined second edition, with an appendix by Karl Rubin, Graduate Texts in Mathematics, **121**, (1990).

[Sa]     A. Saikia, *A simple proof of a lemma of Coleman*, Math. Proc. Cambridge Philos. Soc. **130** (2001), 209–220.

[Se]     J.-P. Serre, *Sur le résidu de la fonction zêta p-adique d'un corps de nombres*, C.R. Acad. Sc. Sér. A **287** (1978), 183–188.

[Se1]    J.-P. Serre, *Classes des corps cyclotomiques (d'après K. Iwasawa)*. Séminaire Bourbaki Exp. no. **174**, (1958).

[SU]     C. Skinner, E. Urban, paper in preparation.

[Si]    W. Sinnott, *On the Stickelberger ideal and the circular units of a cyclotomic field*, Ann. of Math. **108** (1978), 107–134.

[Th]    F. Thaine, *On the ideal class groups of real abelian number fields*, Ann. of Math. **128** (1988), 1–18.

[Y]    R. Yager, *On two variable p-adic L-functions*, Ann. of Math. **115** (1982), 411–449.

[Wa]    L. Washington, Introduction to Cyclotomic Fields, Graduate Texts in Mathematics **83**, Springer (1982).

[W]    A. Wiles, *The Iwasawa conjecture for totally real fields*, Ann. of Math. **131** (1990), 493–540.

# *Springer* **M***onographs in* **M***athematics*

This series publishes advanced monographs giving well-written presentations of the "state-of-the-art" in fields of mathematical research that have acquired the maturity needed for such a treatment. They are sufficiently self-contained to be accessible to more than just the intimate specialists of the subject, and sufficiently comprehensive to remain valuable references for many years. Besides the current state of knowledge in its field, an SMM volume should also describe its relevance to and interaction with neighbouring fields of mathematics, and give pointers to future directions of research.